ENERGY POLICIES, POLITICS AND PRICES

WORLD BIOFUELS PRODUCTION POTENTIAL

ENERGY POLICIES, POLITICS AND PRICES

Nuclear Power's Role in Generating Electricity
Perry G. Furham
2009. ISBN: 978-1-60741-226-7

Dynamic Noncooperative Game Models for Deregulated Electricity Markets
Jose B. Cruz, Jr. and Xiaohuan Tan
2009. ISBN: 978-1-60741-078-2

Hydrogen Fuel Perspectives
Ian S. Rubio (Editor)
2009. ISBN: 978-1-60692-444-0

Power Plant Characteristics and Costs
Stan Kaplan
2009. ISBN: 978-1-60741-264-9

OPEC, Oil Prices and LNG
Edward R. Pitt and Christopher N. Leung (Editors)
2009. ISBN: 978-1-60692-897-4
2009. ISBN: 978-1-60876-614-7 (E-book)

Rational Use and Energy Planning: A Thermodynamic and Geographical Approach
Giuseppe Grazzini, Carla Balocco and Giovan Battista Andreani
2009. ISBN: 978-1-60741-350-9

Carbon Dioxide Emissions
James P. Mulligan (Editor)
2010. ISBN: 978-1-60692-229-3
2010. ISBN: 978-1-61728-096-2 (E-book)

Energy Prices: Supply, Demand or Speculation?
John T. Perry (Editor)
2010. ISBN: 978-1-60741-374-5

Strategic Petroleum Reserve
Albert L. Strait (Editor)
2010. ISBN: 978-1-60692-290-3

Federal Energy Management and Government Efficiency Goals
Amelia R. Williams (Editor)
2010. ISBN: 978-1-60692-985-8
2010. ISBN: 978-1-61668-538-6 (E-book)

Worldwide Biomass Potential: Technology Characterizations
R. L. Bain
2010. ISBN: 978-1-60741-267-0

Solar Energy Technologies: From Research to Deployment
Liam G. White (Editor)
2010. ISBN: 978-1-60741-323-3

The Completion of the Oil Era: The Economic Impact
Carlos A. Rossi
2010. ISBN: 978-1-60741-340-0

U.S. Energy and the Environment: An Overview and Comparative Analysis
Roland H. Terrison (Editor)
2010. ISBN: 978-1-61668-017-6
2010. ISBN: 978-1-61668-641-3 (E-book)

Transition to Hydrogen Fuel Cell Vehicles
Selim Koca (Editor)
2010. ISBN: 978-1-60741-806-1

Employment Effects of Transition to a Hydrogen Economy in the U.S.
Michele Auriemma (Editor)
2010. ISBN: 978-1-60741-808-5

World Biofuels Production Potential
Thomas E. Rommer (Editor)
2010. ISBN: 978-1-61668-663-5
2010. ISBN: 978-1-61668-425-9 (E-book)

U.S. Energy: Overview of the Trends, Statistics, Supply and Consumption
Gregor E. Peake (Editor)
2010. ISBN: 978-1-60876-041-1

Ethanol Economics and Ethanol's Impact on Food Prices and Greenhouse Gas Emissions
Darlene E. Marshall (Editor)
2010. ISBN: 978-1-60876-081-7
2010. ISBN: 978-1- 61728-073-3 (E-book)

The Role of Auctions in Emission Allowance Allocations for Greenhouse Gases
Aubrey D. O'Connor (Editor)
2010. ISBN: 978-1-60741-699-9
2010. ISBN: 978-1-61668-652-9 (E-book)

Reducing Greenhouse Gas Emissions
Joseph G. Levitt (Editor)
2010. ISBN: 978-1-60741-890-0
2010. ISBN: 978-1-61668-730-4 (E-book)

Natural Gas Markets and Lessons Learned
E. K. Cho (Editor)
2010. ISBN: 978-1-61668-249-1
2010. ISBN: 978-1-61668-429-7 (E-book)

Energy Efficiency through Combined Heat and Power or Cogeneration
David H. Thomas (Editor)
2010. ISBN: 978-1-61668-341-2
2010. ISBN: 978-1-61668-432-7 (E-book)

The Smart Grid and Electric Power Transmission
Caitlin G. Elsworth (Editor)
2010. ISBN: 978-1-61668-223-1

Combined Heat and Power - Analysis of Various Markets
Jordan A. Cory
2010. ISBN: 978-1-60741-269-4
2010. ISBN: 978-1-61668-377-1 (E-book)

ENERGY POLICIES, POLITICS AND PRICES

WORLD BIOFUELS PRODUCTION POTENTIAL

THOMAS E. ROMMER
EDITOR

Nova Science Publishers, Inc.
New York

Copyright © 2010 by Nova Science Publishers, Inc.

All rights reserved. No part of this book may be reproduced, stored in a retrieval system or transmitted in any form or by any means: electronic, electrostatic, magnetic, tape, mechanical photocopying, recording or otherwise without the written permission of the Publisher.

For permission to use material from this book please contact us:
Telephone 631-231-7269; Fax 631-231-8175
Web Site: http://www.novapublishers.com

NOTICE TO THE READER

The Publisher has taken reasonable care in the preparation of this book, but makes no expressed or implied warranty of any kind and assumes no responsibility for any errors or omissions. No liability is assumed for incidental or consequential damages in connection with or arising out of information contained in this book. The Publisher shall not be liable for any special, consequential, or exemplary damages resulting, in whole or in part, from the readers' use of, or reliance upon, this material. Any parts of this book based on government reports are so indicated and copyright is claimed for those parts to the extent applicable to compilations of such works.

Independent verification should be sought for any data, advice or recommendations contained in this book. In addition, no responsibility is assumed by the publisher for any injury and/or damage to persons or property arising from any methods, products, instructions, ideas or otherwise contained in this publication.

This publication is designed to provide accurate and authoritative information with regard to the subject matter covered herein. It is sold with the clear understanding that the Publisher is not engaged in rendering legal or any other professional services. If legal or any other expert assistance is required, the services of a competent person should be sought. FROM A DECLARATION OF PARTICIPANTS JOINTLY ADOPTED BY A COMMITTEE OF THE AMERICAN BAR ASSOCIATION AND A COMMITTEE OF PUBLISHERS.

LIBRARY OF CONGRESS CATALOGING-IN-PUBLICATION DATA

Available Upon Request

ISBN : 978-1-61668-663-5

Published by Nova Science Publishers, Inc. ✟ New York

CONTENTS

Preface		ix
Chapter 1	Ethanol Imports and the Caribbean Basin Initiative *Brent D. Yacobucci*	1
Chapter 2	Selected Issues Related to An Expansion of the Renewable Fuel Standard (RFS) *Brent D. Yacobucci and Tom Capehart*	9
Chapter 3	World Biofuels Production Potential: Understanding the Challenges to Meeting the U.S. Renewable Fuel Standard *Office of Policy Analysis*	37
Chapter Sources		109
Index		111

PREFACE

This new book examines the production potential of biofuels in the world. World biofuels production was assessed over the 2010 to 2030 timeframe using scenarios covering a range of U.S. policies (tax credits, tariffs, and regulations) as well as oil prices, feedstock availability, and a global CO_2 price.

Chapter 1 - Fuel ethanol consumption has grown significantly in the past several years, and it will continue to grow with the establishment of a renewable fuel standard (RFS) in the Energy Policy Act of 2005 (P.L. 109-58) and the expansion of that RFS in the Energy Independence and Security Act of 2007 (P.L. 110-140). This standard requires U.S. transportation fuels to contain a minimum amount of renewable fuel, including ethanol.

Most of the U.S. market is supplied by domestic refiners producing ethanol from American corn. However, imports play a small but growing role in the U.S. market. One reason for the relatively small role is a 2.5% ad valorem tariff and (more significantly) a 54-cent-per-gallon added duty on imported ethanol. These duties offset an economic incentive of 51 cents per gallon for the use of ethanol in gasoline. However, to promote development and stability in the Caribbean region and Central America, the Caribbean Basin Initiative (CBI) allows the imports of most products, including ethanol, duty-free. While many of these products are produced in CBI countries, ethanol entering the United States under the CBI is generally produced elsewhere and reprocessed in CBI countries for export to the United States. The U.S.- Central America Free Trade Agreement (CAFTA) would maintain this duty-free treatment and set specific allocations for imports from Costa Rica and El Salvador. Duty-free treatment of CBI ethanol has raised concerns, especially as the market for ethanol has the potential for dramatic expansion under P.L. 109-5 8 and P.L. 110-140.

In the United States, fuel ethanol is largely domestically produced. A value-added product of agricultural commodities, mainly corn, it is used as a gasoline additive and as an alternative to gasoline. To promote its use, ethanol-blended gasoline is granted a significant tax incentive. However, this incentive does not recognize point of origin, and there is a duty on most imported fuel ethanol to offset the exemption. But a limited amount of ethanol may be imported under the Caribbean Basin Initiative (CBI) duty-free, even if most of the steps in the production process were completed in other countries. This duty-free import of ethanol has raised concerns, especially as U.S. demand for ethanol has been growing. Further, duty-free imports from these countries, especially Costa Rica and El Salvador, have played a role in the development of the U.S.-Central America Free Trade Agreement (CAFTA).

Chapter 2 - High petroleum and gasoline prices, concerns over global climate change, and the desire to promote domestic rural economies have greatly increased interest in biofuels as an alternative to petroleum in the U.S. transportation sector. Biofuels, most notably corn ethanol, have grown significantly in the past few years as a component of U.S. motor fuel supply. Ethanol, the most commonly used biofuel, is blended in more than half of all U.S. gasoline (at the 10% level or lower in most cases). However, current biofuels supply of 6.8 billion gallons only represents about 4% of total vehicle fuel demand.

The Energy Independence and Security Act of 2007 (EISA, P.L. 110-140) requires ever-larger amounts of biofuels produced from feedstocks other than corn starch, including sugarcane, oil crops, and cellulose, and promotes the development of these fuels. EISA requires the use of 36 billion gallons of renewable fuels annually in 2022, of which only 15 billion gallons can be ethanol from corn starch. The remaining 21 billion gallons are to be so-called "advanced biofuels." The previous RFS in the Energy Policy Act of 2005 (P.L. 109-58) required the use of only 7.5 billion gallons in 2012, increasing to an expected 8.6 billion gallons in 2022, of which only 250 million gallons of cellulosic biofuels would be required.

Although EISA has set the goal of significantly expanding biofuels supply and use in the coming decades, questions remain about the ability of the U.S. biofuels industry to meet the rapidly increasing mandate. Current U.S. biofuels supply relies almost exclusively on ethanol produced from Midwest corn. During the 2008 crop year, 31% of the U.S. corn crop is projected to be used for ethanol production.

Due to the concerns with significant expansion in corn-based ethanol supply, interest has grown in expanding the market for biodiesel produced from soybeans and other oil crops. However, a significant increase in U.S. biofuels would likely require a movement away from food and grain crops as feedstocks. Other biofuels feedstock sources, including cellulosic biomass, are promising, but technological barriers make their future uncertain.

Issues facing the U.S. biofuels industry include potential agricultural feedstock supplies, the associated market and environmental effects of a major shift in U.S. agricultural production; the energy consumed to grow feedstocks and process them into fuel, and barriers to expanded infrastructure needed to deliver more and more biofuels to the market. Key questions are whether a renewable fuel mandate is the most effective policy to promote the above goals, if government intervention in the industry is appropriate, and, if so, what level is appropriate. This report outlines some of the current supply issues facing biofuels industries, including implications for agricultural feedstocks, infrastructure concerns, energy supply for biofuels production, and fuel price uncertainties.

This report supersedes CRS Report RL34265, *Selected Issues Related to an Expansion of the Renewable Fuel Standard (RFS)*, by Brent D. Yacobucci and Tom Capehart.

Chapter 3 - This study by the U.S. Department of Energy (DOE) estimates the worldwide potential to produce biofuels including biofuels for export. It was undertaken to improve our understanding of the potential for imported biofuels to satisfy the requirements of Title II of the 2007 Energy Independence and Security Act (EISA) in the coming decades[1]. Many other countries' biofuels production and policies are expanding as rapidly as ours. Therefore, authors modeled a detailed and up-to-date representation of the amount of biofuel feedstocks that are being and can be grown, current and future biofuels production capacity, and other factors relevant to the economic competitiveness of worldwide biofuels production, use, and trade.

The Oak Ridge National Laboratory (ORNL) identified and prepared feedstock data for countries that were likely to be significant exporters of biofuels to the U.S. The National Renewable Energy Laboratory (NREL) calculated conversion costs by conducting material flow analyses and technology assessments on biofuels technologies. Brookhaven National Laboratory (BNL) integrated the country specific feedstock estimates and conversion costs into the global Energy Technology Perspectives (ETP) MARKAL (MARKet ALlocation) model. The model uses least-cost optimization to project the future state of the global energy system in five year increments. World biofuels production was assessed over the 2010 to 2030 timeframe using scenarios covering a range U.S. policies (tax credits, tariffs, and regulations), as well as oil prices, feedstock availability, and a global CO_2 price.

All scenarios include the full implementation of existing U.S. and selected other countries' biofuels' policies (Table 4). For the U.S., the most important policy is the EISA Title II Renewable Fuel Standard (RFS). It progressively increases the required volumes of renewable fuel used in motor vehicles (Appendix B). The RFS requires 36 billion (B) gallons (gal) per year of renewable fuels by 2022[2]. Within the mandate, amounts of advanced biofuels, including biomass-based diesel and cellulosic biofuels, are required beginning in 2009. Imported renewable fuels are also eligible for the RFS. Another key U.S. policy is the $1.01 per gal tax credit for producers of cellulosic biofuels enacted as part of the 2008 Farm Bill[3]. This credit, along with the DOE's research, development and demonstration (RD&D) programs, are assumed to enable the rapid expansion of U.S. and global cellulosic biofuels production needed for the U.S. to approach the 2022 RFS goal[4]. While the Environmental Protection Agency (EPA) has yet to issue RFS rules to determine which fuels would meet the greenhouse gas (GHG) reduction and land use restrictions specified in EISA, we assume that cellulosic ethanol, biomass-to-liquid fuels (BTL), sugar-derived ethanol, and fatty acid methyl ester biodiesel would all meet the EISA advanced biofuel requirements. We also assume that enough U.S. corn ethanol would meet EISA's biofuel requirements or otherwise be grandfathered under EISA to reach 15 B gal per year.

In: World Biofuels Production Potential
Editor: Thomas E. Rommer

ISBN: 978-1-61668-663-5
© 2010 Nova Science Publishers, Inc.

Chapter 1

ETHANOL IMPORTS AND THE CARIBBEAN BASIN INITIATIVE

Brent D. Yacobucci

SUMMARY

Fuel ethanol consumption has grown significantly in the past several years, and it will continue to grow with the establishment of a renewable fuel standard (RFS) in the Energy Policy Act of 2005 (P.L. 109-58) and the expansion of that RFS in the Energy Independence and Security Act of 2007 (P.L. 110-140). This standard requires U.S. transportation fuels to contain a minimum amount of renewable fuel, including ethanol.

Most of the U.S. market is supplied by domestic refiners producing ethanol from American corn. However, imports play a small but growing role in the U.S. market. One reason for the relatively small role is a 2.5% ad valorem tariff and (more significantly) a 54-cent-per-gallon added duty on imported ethanol. These duties offset an economic incentive of 51 cents per gallon for the use of ethanol in gasoline. However, to promote development and stability in the Caribbean region and Central America, the Caribbean Basin Initiative (CBI) allows the imports of most products, including ethanol, duty-free. While many of these products are produced in CBI countries, ethanol entering the United States under the CBI is generally produced elsewhere and reprocessed in CBI countries for export to the United States.

The U.S.- Central America Free Trade Agreement (CAFTA) would maintain this duty-free treatment and set specific allocations for imports from Costa Rica and El Salvador. Duty-free treatment of CBI ethanol has raised concerns, especially as the market for ethanol has the potential for dramatic expansion under P.L. 109-5 8 and P.L. 110-140.

In the United States, fuel ethanol is largely domestically produced. A value-added product of agricultural commodities, mainly corn, it is used as a gasoline additive and as an alternative to gasoline. To promote its use, ethanol-blended gasoline is granted a significant tax incentive. However, this incentive does not recognize point of origin, and there is a duty on most imported fuel ethanol to offset the exemption. But a limited amount of ethanol may be imported under the Caribbean Basin Initiative (CBI) duty-free, even if most of the steps in the production process were completed in other countries. This duty-free import of ethanol has raised concerns, especially as U.S. demand for ethanol has been growing. Further, duty-free imports from these countries, especially Costa Rica and El Salvador, have played a role in the development of the U.S.-Central America Free Trade Agreement (CAFTA).

Fuel Ethanol

Ethanol is an alcohol fuel produced from the fermentation of simple sugars.[1] Most ethanol in the United States is produced from corn. In other countries, sugarcane or other plants are common feedstocks. In the United States, the increased demand for corn leads to higher revenues for U.S. corn farmers. Ethanol is usually blended in gasoline (a mixture called "gasohol") to increase octane, improve combustion, and extend gasoline stocks. Currently, about 3% to 5% of total U.S. gasoline demand is actually met by ethanol, and roughly half of U.S. gasoline contains some ethanol.

U.S. ethanol is generally produced and consumed in the Midwest, close to where the corn feedstock is produced. The main steps to ethanol production are as follows:

- The feedstock (e.g., corn) is processed to separate fermentable sugars.
- Yeast is added to ferment the sugars.
- The resulting alcohol is distilled.
- Finally, the distilled alcohol is dehydrated to remove any remaining water.

This final step — dehydration — is at the heart of the issue over ethanol imports from the CBI, as discussed below.

Ethanol Imports

According to the United States International Trade Commission, the majority of all fuel ethanol imports to the United States came through CBI countries between 1999 and 2003 (see **Figure 1**).[2] In 2004, imports from Brazil to the United States grew dramatically, but in 2005, CBI imports again represented more than half of all U.S. ethanol imports. With an increase in ethanol demand in 2006 due to voluntary elimination of MTBE — a competitor for ethanol in gasoline blending — imports grew dramatically, roughly quadrupling imports in any previous year.[3] Most of this increase was in direct imports from Brazil. Historically, imports have played a relatively small role in the U.S. ethanol market. Total ethanol consumption in 2005 was approximately 3.9 billion gallons, whereas imports totaled 135 million gallons, or about 4%. Imports from the CBI totaled approximately 2.6%. In 2006, total imports represented roughly 13% of the 5.0 billion gallons consumed in 2006; ethanol from CBI countries represented roughly 3.4%. In 2007, total imports represented roughly 6% of U.S. consumption (6.8 billion gallons); ethanol from CBI countries represented roughly 3.6%.

One reason for limited imports — even though, in some cases, production costs for ethanol in foreign countries are significantly lower than in the United States — is a mostfavored-nation tariff of 2.5% and an added duty of 54 cents per gallon.[4] In many cases, this tariff negates lower production costs in other countries. For example, by some estimates, Brazilian production costs have been roughly 50% lower than in the United States.[5] A key motivation for the establishment of the tariff was to offset a tax incentive for ethanol-blended gasoline ("gasohol").[6] This incentive is currently valued at 51 cents per gallon of pure ethanol used in blending. Unless imports enter the United States duty-free, the tariff effectively negates the incentive for those imports. With U.S. wholesale ethanol prices ranging from roughly $1.50 to $2.50 per gallon for most of the time between January 2006 to March 2008, the tariff has presented a significant barrier to imports.[7] However, during the voluntary phaseout of MTBE, there was a significant spike in wholesale prices between April 2006 and September 2006, with wholesale prices nearing $6.00 per gallon in some markets during the summer of 2006.[8] This runup in prices significantly improved the profitability of importing ethanol, regardless of the duty.

Ethanol and the CBI

As Congress noted in the Customs and Trade Act of 1990, the Caribbean Basin Initiative (CBI) was established in 1983 to promote "a stable political and economic climate in the Caribbean region."[9] As part of the initiative, duty-free status is granted to a large array of products from beneficiary countries, including fuel ethanol under certain conditions. If produced from at least 50% local feedstocks (e.g., ethanol produced from sugarcane grown in the CBI beneficiary countries), ethanol may be imported duty-free.[10] If the local feedstock content is lower, limitations apply on the quantity of duty-free ethanol. Nevertheless, up to 7% of the U.S. market may be supplied duty-free by CBI ethanol containing no local feedstock.[11] In this case, hydrous ("wet") ethanol produced in other countries, historically Brazil or European countries, can be shipped to a dehydration plant in a CBI country for reprocessing.[12] After the ethanol is dehydrated, it is imported duty-free into the United States. Currently, imports of dehydrated ethanol under the CBI are far below the 7% cap (approximately 3% in 2006). For 2006, the cap was about 270 million gallons,[13] whereas about 170 million gallons were imported under the CBI in that year.[14]

Dehydration plants are currently operating in Jamaica, Costa Rica, El Salvador, Trinidad and Tobago, and the U.S. Virgin Islands.[15] Jamaica and Costa Rica were the two largest exporters of fuel ethanol to the United States from 1999 to 2003. (In 2004 and 2006, direct imports from Brazil exceeded imports from all other countries combined.)[16]

Despite criticisms in the United States, new dehydration facilities began production in Trinidad and Tobago in 2005[17] and the U.S. Virgin Islands in 2007.

Duty-free ethanol imports have also played a role in discussions regarding the U.S.- Central America Free Trade Agreement (CAFTA).[18] Under this agreement signed by the Bush Administration and the participating countries, specific allocations (of the 7% duty- free cap for CBI ethanol) are set aside for Costa Rica and El Salvador. These allocations effectively limit the amount of fuel that other CBI countries can import duty-free. Costa Rica's allocation is 31 million gallons per year, while El Salvador was granted an initial allocation of approximately 6.6 million gallons per year, increasing by roughly 1.3 million gallons in each subsequent year. However, El Salvador's allocation may not exceed 10% of the total CBI allocation (or 0.7% of the U.S. market). The agreement was signed on May 28, 2004. Congress approved the agreement in 2005, and implementing legislation was signed by President

Bush on August 2, 2005 (P.L. 109-53). As both countries exceeded their allocations in 2005, 2006, and 2007, the ultimate effects of the allocations is unclear.

Growing U.S. Ethanol Market

The U.S. ethanol market has grown dramatically over the past several years. Between 1990 and 2007, U.S. ethanol consumption increased from about 900 million gallons per year to 6.8 billion gallons per year. Much of this growth has resulted from Clean Air Act requirements that gasoline in areas with the worst ozone pollution contain an oxygenate, such as ethanol, and the establishment of a renewable fuel standard (RFS) in the Energy Policy Act of 2005 (P.L. 109-5 8). The RFS required that gasoline sold in the United States contain a renewable fuel, such as ethanol. The mandate required 4.0 billion gallons of renewable fuel in 2006, increasing to 7.5 billion gallons in 2012. The Energy Independence and Security Act of 2007 (P.L.1 10-140) expanded the RFS to 9.0 billion gallons in 2008, increasing to 36 billion gallons in 2022. In addition, the expanded RFS specifically requires the use of an increasing amount of "advanced biofuels" — biofuels produced from feedstocks other than corn starch (including sugar cane ethanol). While domestic producers anticipate greater demand for their product under the RFS, they are also concerned that duty-free ethanol imports through the CBI could dramatically increase, to their detriment.

Duty Drawback

In addition to the concerns over imports of duty-free ethanol from CBI countries, there is growing concern that a large portion of ethanol otherwise subject to the duties is being imported duty-free through a "manufacturing drawback."[19] If a manufacturer imports an intermediate product then exports the finished product or a similar product, that manufacturer may be eligible for a refund (drawback) of up to 99% of the duties paid. There are special provisions for the production of petroleum derivatives.[20] In the case of fuel ethanol, the imported ethanol is used as a blending component in gasoline, and jet fuel (considered a like commodity) is exported to qualify for the drawback.[21] Some critics estimate that as much as 75% or more of the duties were eligible for the drawback in 2006. Therefore, critics question the effectiveness of the ethanol duties and the CBI exemption.

Congressional Action

Some Members of Congress have expressed concern over duty-free imports of dehydrated ethanol that originates in Brazil or other countries. Therefore, there is growing interest from some Members of Congress to eliminate the CBI exemption and/or modify the manufacturing drawback for petroleum products.

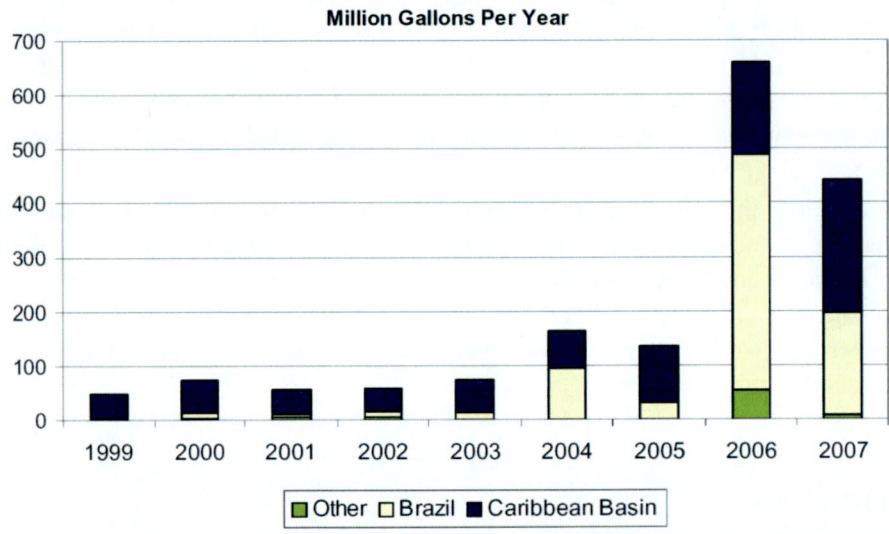

Source: U.S. International Trade Commission (USITC), Interactive Tariff and Trade Data Web, at [http://dataweb.usitc.gov], accessed March 9, 2006, and USITC, U.S. Imports of Fuel Ethanol, by Source 1996-2007, updated February 2008.

Figure 1. Annual Ethanol Imports to the United States

Although some stakeholders are concerned over increased ethanol imports and their effect on the U.S. industry, others believe that tariffs on imported ethanol should be eliminated entirely. They argue that increased use of ethanol, regardless of its origin, would further displace gasoline consumption. They also argue that inexpensive imported ethanol would help mitigate any fuel price increases from the renewable fuels standard.

Conclusion

With growing demand for ethanol, there is increased interest in foreign imports. Because ethanol from CBI countries is granted duty-free status, there is the possibility that imports of dehydrated ethanol will grow because of this avenue provided in the law. While CBI countries have not yet reached their quota for ethanol refined in other countries and dehydrated in the Caribbean, CBI imports have increased over the past few years, and may exceed the quota in future years. CBI imports have the potential to increase significantly over the next few years, especially as the domestic market grows under the renewable fuels standard. In addition, the manufacturing drawback could provide another avenue for duty-free ethanol imports directly from Brazil and other countries.

Low-cost ethanol imports could have an advantage over domestically produced ethanol, which could affect the U.S. ethanol industry and American corn growers. However, the U.S. ethanol industry has grown significantly in the past several years, and will likely continue to grow regardless of the level of imports.

End Notes

[1] For more information on ethanol, see CRS Report RL30369, *Fuel Ethanol: Background and Public Policy Issues*, by Brent D. Yacobucci.

[2] It should be noted that between 1999 and 2003, Saudi Arabia was the largest exporter to the United States of ethanol. However, this ethanol is synthetic (produced from fossil fuels) and does not qualify for the tax incentives for ethanol-blended fuel. Therefore, ethanol from Saudi Arabia is used as an industrial feedstock and is subject to different tariff treatment than fuel ethanol.

[3] For more information on the MTBE phaseout, see CRS Report RL31361, *"Boutique Fuels" and Reformulated Gasoline: Harmonization of Fuel Standards*, by Brent D. Yacobucci.

[4] Technically, the tariff is 14.27 cents per liter, which is equal to 54 cents per gallon.

[5] "NCGA's Adams Addresses World Energy Crisis at ACE Meeting," *NCGA News*, August 16, 2004; Kevin Diaz, "Cargill Takes Heat Over Ethanol Import Plan," *Star Tribune*, July 2, 2004.

[6] U.S. General Accounting Office, *Fuel Ethanol: Imports from Caribbean Basin Initiative Countries*, April 1989. For more information on the excise tax exemption, see CRS Report 98-435, *Alcohol Fuels Tax Incentives*, by Salvatore Lazzari.

[7] Chemical Week Associates, "Octane Week Price Report," *Octane Week*, various issues, January 2006 to March 2006, and Chicago Board of Trade, *Ethanol Derivatives, updated through January, 2008*, Chicago, February 13, 2008.

[8] Chicago Board of Trade, *op. cit.*

[9] P.L. 101-382, §202; 19 U.S.C. 2701 note: congressional findings.

[10] P.L. 99-514, §423; 19 U.S.C. 2703 note: ethyl alcohol and mixtures thereof for fuel use.

[11] Ibid.

[12] U.S. House of Representatives, Committee on Ways and Means, *Hearing on Fuel Ethanol Imports from Caribbean Basin Initiative Countries*, April 25, 1989.

[13] *69 Federal Register 76956.*

[14] The quota for a given year is calculated based on 7% of U.S. consumption in the preceding year. Therefore, as U.S. consumption is growing, the quota represents somewhat less than 7% of total U.S. consumption in that year.

[15] Petrojam, Ltd., *Petrojam Ethanol Limited - Alcohol Sources.* U.S. International Trade Commission (USITC), *U.S. Imports of Fuel Ethanol, by Source 1996-200 7*, February, 2008

[16] USITC, *Interactive Tariff and Trade Data Web*, at [http://dataweb.usitc.gov]. March 9, 2006.

[17] This project has received particular scrutiny from some critics because its construction was financed through a loan insured by the U.S. Export-Import Bank.

[18] For more information on CAFTA, see CRS Report RL3 1870, *The Dominican Republic-Central America-United States Free Trade Agreement (CAFTA-DR)*, by J. F. Hornbeck.

[19] For more information on drawbacks, see U.S. Customs Service, *Drawback: A Refund for Certain Exports*, Washington, February 2002.

[20] 19 U.S.C. 1313(p).

[21] Peter Rhode, "Senate Finance May Take Up Drawback Loophole As Part Of Energy Bill," *Energy Washington Week*, April 18, 2007.

In: World Biofuels Production Potential
Editor: Thomas E. Rommer

ISBN: 978-1-61668-663-5
© 2010 Nova Science Publishers, Inc.

Chapter 2

SELECTED ISSUES RELATED TO AN EXPANSION OF THE RENEWABLE FUEL STANDARD (RFS)

Brent D. Yacobucci and Tom Capehart

SUMMARY

High petroleum and gasoline prices, concerns over global climate change, and the desire to promote domestic rural economies have greatly increased interest in biofuels as an alternative to petroleum in the U.S. transportation sector. Biofuels, most notably corn ethanol, have grown significantly in the past few years as a component of U.S. motor fuel supply. Ethanol, the most commonly used biofuel, is blended in more than half of all U.S. gasoline (at the 10% level or lower in most cases). However, current biofuels supply of 6.8 billion gallons only represents about 4% of total vehicle fuel demand.

The Energy Independence and Security Act of 2007 (EISA, P.L. 110-140) requires ever-larger amounts of biofuels produced from feedstocks other than corn starch, including sugarcane, oil crops, and cellulose, and promotes the development of these fuels. EISA requires the use of 36 billion gallons of renewable fuels annually in 2022, of which only 15 billion gallons can be ethanol from corn starch. The remaining 21 billion gallons are to be so-called "advanced biofuels." The previous RFS in the Energy Policy Act of 2005 (P.L. 109-58) required the use of only 7.5 billion gallons in 2012, increasing to an expected 8.6 billion gallons in 2022, of which only 250 million gallons of cellulosic biofuels would be required.

Although EISA has set the goal of significantly expanding biofuels supply and use in the coming decades, questions remain about the ability of the U.S. biofuels industry to meet the rapidly increasing mandate. Current U.S. biofuels supply relies almost exclusively on ethanol produced from Midwest corn. During the 2008 crop year, 31% of the U.S. corn crop is projected to be used for ethanol production.

Due to the concerns with significant expansion in corn-based ethanol supply, interest has grown in expanding the market for biodiesel produced from soybeans and other oil crops. However, a significant increase in U.S. biofuels would likely require a movement away from food and grain crops as feedstocks. Other biofuels feedstock sources, including cellulosic biomass, are promising, but technological barriers make their future uncertain.

Issues facing the U.S. biofuels industry include potential agricultural feedstock supplies, the associated market and environmental effects of a major shift in U.S. agricultural production; the energy consumed to grow feedstocks and process them into fuel, and barriers to expanded infrastructure needed to deliver more and more biofuels to the market. Key questions are whether a renewable fuel mandate is the most effective policy to promote the above goals, if government intervention in the industry is appropriate, and, if so, what level is appropriate. This report outlines some of the current supply issues facing biofuels industries, including implications for agricultural feedstocks, infrastructure concerns, energy supply for biofuels production, and fuel price uncertainties.

This report supersedes CRS Report RL34265, *Selected Issues Related to an Expansion of the Renewable Fuel Standard (RFS)*, by Brent D. Yacobucci and Tom Capehart.

INTRODUCTION

High petroleum and gasoline prices, concerns over global climate change, and the desire to promote domestic rural economies have raised interest in biofuels as an alternative to petroleum in the U.S. transportation sector. Biofuels, most notably corn-based ethanol, have grown significantly in the past few years as a component of U.S. motor fuels. More than half of all U.S. gasoline contains some ethanol (mostly blended at the 10% level or lower).

However, current supply represents only about 5% of annual gasoline demand on a volume basis, and only about 3% on an energy basis. In 2007, the United States consumed roughly 6.8 billion gallons of ethanol; this 6.8 billion gallons was blended into roughly 136 billion gallons of gasoline.

In his 2007 State of the Union Address, President Bush expressed support for expanding biofuels supply significantly in the coming decades. President Bush proposed expanding consumption from 5 billion gallons in 2007 to 35 billion gallons in 2017. Although this proposal included not just biofuels but alternative fuels in general (including fuels from coal or natural gas), it suggested a significant growth in biofuels production over the next 10 years. Legislative proposals in the 110th Congress would have required significant expansion of biofuels production in the coming decades; some proposals would have required 30 billion gallons of biofuels alone by 2030 or 60 billion gallons by 2050. The Energy Independence and Security Act of 2007 (EISA, P.L. 110-140) contains a renewable fuel standard (RFS) that requires the use of 36 billion gallons in 2022, including 21 billion gallons of advanced biofuels."[1] The law limits ethanol from corn starch under the RFS to 15 billion gallons beginning in 2015.

Current U.S. biofuels supply relies almost exclusively on ethanol produced from Midwest corn (**Table 1**). Other fuels that play a smaller role include ethanol from Brazilian sugar, biodiesel from U.S. soybeans, and ethanol from U.S. sorghum. A significant increase in U.S. biofuels would likely require a movement away from food and grain crops. For example, U.S. ethanol production in 2008 is projected to consume roughly 31% of the U.S. corn crop. Under the expanded RFS, the 15 billion gallon corn ethanol cap would place a call on nearly half the volume of corn produced in 2008. Corn (and other grains) have myriad other uses, and such a shift could have drastic consequences for most agricultural inputs:

- grains—because corn would compete with other grains for land;
- livestock—because the cost of animal feed would likely increase; and
- land—because total harvested acreage would likely increase.

In addition to agricultural effects, such an increase in corn-based ethanol would likely have other effects, including:

- fuel costs—because biofuels tend to be more expensive than petroleum fuels;

- energy supply—because natural gas is a key input into corn production; and
- the environment—because the expansion of corn-based ethanol production raises many environmental questions.

These concerns are discussed below.

BIOFUELS DEFINED

Any fuel produced from biological materials (e.g., food crops, agricultural residues, municipal waste) is generally referred to as a biofuel." More specifically, the term generally refers to liquid transportation fuels. As stated above, the most significant biofuel in the United States is ethanol produced from corn. Approximately 6.5 billion gallons of ethanol were produced in the United States in 2007,[2] mostly from corn. Other domestic feedstocks for ethanol include grain sorghum and sweet sorghum; imported ethanol (435 million gallons in 2007) is usually produced from sugar cane in Brazil. Ethanol is generally blended into gasoline at the 10% level (E10) or lower. It can be used in purer forms such as E85 (85% ethanol and 15% gasoline) in vehicles specially designed for its use, although E85 represents less than 1% of U.S. ethanol consumption.

Due to concerns over the significant expansion in corn-based ethanol supply, interest has grown in expanding the market for biodiesel (a diesel substitute produced from vegetable and animal oils, mainly soybean oil) and spurring the development of motor fuels produced from cellulosic materials including grasses, trees, and agricultural and municipal wastes. However, all of these so-called advanced biofuels technologies are currently even more expensive than corn-based ethanol (with the exception of ethanol produced from Brazilian sugarcane).

In addition to expanding domestic production of biofuels, there is some interest in expanding imports of sugar-based ethanol from Brazil and other countries. However, ethanol from Brazil is currently subject to a $0.54 per gallon tariff that in most years is a significant barrier to direct Brazilian imports.[3] Some Brazilian ethanol can be brought into the United States duty free if it is dehydrated (reprocessed) in Caribbean Basin Initiative (CBI) countries.[4] Up to 7% of the U.S. ethanol market could be supplied duty-free in this fashion; historically, however, ethanol dehydrated in CBI countries has only represented about 2% of the total U.S. market.

After ethanol, biodiesel is the next most significant biofuel in the United States, although 2007 U.S. production is estimated at only 491 million gallons,[5] compared to roughly 45 billion gallons of on-road diesel fuel in the same year.[6] Other biofuels with the potential to play a role in the U.S. market include diesel fuel substitutes and ethanol produced from various biomass feedstocks containing cellulose. However, these cellulosic biofuels are currently prohibitively expensive relative to conventional ethanol and biodiesel. Other potential biofuels include other alcohols (e.g., methanol and butanol) produced from biomass.

This report outlines some of the current issues related to the RFS established in the Energy Policy Act of 2005 (P.L. 109-58) and expanded in the EISA of 2007, including implications for agricultural feedstocks, infrastructure constraints, environmental concerns, energy supply issues, and fuel prices.

Table 1. U.S. Production of Biofuels from Various FeedstocksBiofuels Defined

Fuel	Feedstock	U.S. Production in 2007
Ethanol	Corn	6.5 billion gallons
	Sorghum	less than 100 million gallons
	Cane sugar	No production (450 million gallons imported from Brazil and Caribbean countries)
	Cellulose	No production (one demonstration plant in Canada)
Biodiesel	Soybean oil	approximately 470 million gallons
	Other vegetable oils	less than 10 million gallons
	Recycled grease	less than 10 million gallons
	Cellulose	No production
Methanol	Cellulose	No production
Butanol	Cellulose other biomass	No production

Source: Renewable Fuels Association; National Biodiesel Board; CRS analysis.

The Renewable Fuel Standard (RFS)

RFS in the Energy Independence and Security Act of 2007 (P.L. 110-140)

Section 202 of EISA requires the use of 9 billion gallons of renewable fuels in 2008, increasing annually to reach 36 billion gallons in 2022. Previously, the Energy Policy Act of 2005 (P.L. 109-58) required, starting in 2006, the use of 4.0 billion gallons of renewable fuels, increasing to 7.5 billion in 2012. Beginning in 2015, only 15 billion gallons can be ethanol from corn starch. Any additional volume is not credited toward the annual mandate under the RFS. The remaining 21 billion gallons in 2022 are to be so-called "advanced biofuels." Currently, production of advanced biofuels is limited to ethanol derived from sugar and biodiesel. Although the RFS has been called an ethanol mandate, there is no explicit requirement to use ethanol. Although there are specific requirements for the use of biodiesel and other renewable fuels, it is expected that, in the early years, the vast majority of the RFS will be met using ethanol produced from corn starch.

This report examines the specific issues regarding the implementation of the expanded RFS contained in EISA, but does not address the broader public policy issue surrounding how best to support U.S. energy policy.

RFS as Public Policy

The expansion in the RFS could have significant policy implications. Issues include questions of energy/petroleum security, pollutant and greenhouse gas emissions, agricultural commodity and food market effects, land use and conservation, and infrastructure costs. Proponents of mandated biofuels use respectively claim that an RFS would promote the general public interest on several different policy fronts, while opponents disagree. For example, supporters of an RFS claim it would serve several public policy interests including:

- reduced investment risk by guaranteeing demand for a projected period (such risk would otherwise keep significant investment capital on the sidelines);
- enhanced energy security via the production of liquid fuel from a renewable domestic source resulting in decreased reliance on

imported fossil fuels (the U.S. currently imports over half of its petroleum, two-thirds of which is consumed by the transportation sector); and
- enhanced environmental benefits (most biofuels are non-toxic, biodegradable, use renewable resources, etc.).

Critics of an RFS have taken issue with many specific aspects of biofuels production and use, but a general public policy criticism of the RFS is that, by picking the winner," policymakers may exclude or retard the development of other, potentially more preferable alternative energy sources.[7] They contend that biofuels are given a huge advantage via billions of dollars of annual subsidies which distort investment markets by redirecting venture capital and other investment dollars away from competing alternative energy sources. Instead, these critics have argued for a more "technology neutral" policy such as a carbon tax, a cap-and-trade system of carbon credits, or a floor price on imported petroleum.

The Expanded RFS Defined

The expanded RFS includes all motor fuel, as well as heating oil (**Table 2**). It reaches 13.2 billion gallons (bgal.) in 2012 (compared with the previous RFS of 7.5 bgal.); 15 bgal. by 2015; and 36 bgal. in 2022. However, the corn based ethanol share of the expanded RFS is capped at 15 bgal. in 2015, and all subsequent annual increases are to be derived entirely from advanced biofuels— defined as biofuels derived from feedstocks other than corn starch. The advanced biofuels volume under the RFS reaches 21 bgal. by 2022.

The expanded RFS requires that renewable fuels produced in facilities that commence operation after enactment must achieve at least a 20% reduction in life-cycle greenhouse gas emissions relative to gasoline. This requirement rises to 50% for advanced biofuels, and 60% for cellulosic biofuels.

The RFS as amended in EISA involves two distinct components—a corn-starch-ethanol RFS and a non-corn-starch-ethanol RFS—that are best analyzed separately because the various supply and demand factors affecting their development also are fairly distinct.

Table 2. EISA 2007 Expansion of the Renewable Fuel Standard

Year	Previous RFS in EPAct of 2005 (billion gallons)	Biofuel mandate for motor fuel, home heating oil, and boiler fuel (billion gallons)	Portion to be from advanced biofuels (i.e., not corn starch) (billion gallons)	Cap on corn starch-derived ethanol (billion gallons)
2006	4.0	4.00	0.00	4.0
2007	4.7	4.70	0.00	4.7
2008	5.4	9.00	0.00	9.0
2009	6.1	11.10	0.60	10.5
2010	6.8	12.95	0.95	12.0
2011	7.4	13.95	1.35	12.6
2012	7.5	15.20	2.00	13.2
2013	7.6 (est.)	16.55	2.75	13.8
2014	7.7 (est.)	18.15	3.75	14.4
2015	7.8 (est.)	20.50	5.50	15.0
2016	7.9 (est.)	22.25	7.25	15.0
2017	8.1 (est.)	24.00	9.00	15.0
2018	8.2 (est.)	26.00	11.00	15.0
2019	8.3 (est.)	28.00	13.00	15.0
2020	8.4 (est.)	30.00	15.00	15.0
2021	8.5	33.00	18.00	15.0

Source: EISA, Section 202.

POTENTIAL ISSUES WITH THE EXPANDED RFS

Overview of Long-Run Corn Ethanol Supply Issues

The U.S. ethanol industry has shown rapid growth in recent years, with national production increasing from 1.8 billion gallons in 2001 to 6.5 billion gallons in 2007. This rapid growth has important consequences for U.S. and international fuel, feed, and food markets.

Corn accounts for about 97% of the feedstocks used in ethanol production in the United States. USDA projects that 3.7 billion bushels of corn (or 31% of the 2008 corn crop) will be used to produce ethanol during the September 2008 to August 2009 corn marketing year.[8] In 2007, U.S. corn production was a record 13.1 billion bushels and production in 2008 is projected pull back to 12.0 billion bushels. As of December 2008, existing U.S. ethanol plant capacity was a reported 10.8 billion gallons per year, with an additional capacity of 2.4 billion gallons under construction or available for expansion.[9]

Thus, total annual U.S. ethanol production capacity in existence or under construction as of December 2008, was 13.2 billion gallons. This potential production capacity exceeds the 13.0 billion gallon supply required in 2010 by EISA. The current pace of plant construction suggests that annual corn-for-ethanol use will likely approach, or possibly exceed, 5 billion bushels by 2010.[10] However, low gasoline prices in late 2008 and the recession's impact on the industry may slow new plant construction and plant expansions.

The ethanol-driven surge in corn demand has been associated with a sharp rise in corn prices. For example, the futures contract for March 2007 corn on the Chicago Board of Trade rose 66% from $2.50 per bushel in September 2006 to a contract high of over $4.16 per bushel in January 2007. Although a record U.S. corn harvest eased upward pressure on corn prices slightly during 2007, by November 2007 prices for 2008 futures contracts were again trading at more than $4.00 per bushel. However, in the summer of 2008, Central Illinois corn prices skyrocketed to a record high $6.55 per bushel.[11] In late 2008, prices fell to below $2.00 per bushel. Volatility in the corn market is largely attributed to the link between the use of corn for both food and fuel. Both USDA and the Food and Agricultural Policy Research Institute (FAPRI) (**Table 3**), in their annual agricultural baseline reports, project corn prices to remain well above $3.00 per bushel through 2016 compared with an average farm price of $2.15 per bushel during the previous 10-year period (1997-2006).

This sharp rise in corn prices owed its origins largely to increasing corn demand spurred by the rapid expansion of corn-based ethanol production capacity in the United States since mid-2006. The rapid growth in ethanol capacity has been fueled by both strong energy prices and a variety of government incentives, regulations, and programs. Major federal incentives include a tax credit of 51 cents to fuel blenders for every gallon of ethanol blended with gasoline; the Renewable Fuel Standard; and the 54 cents per gallon duty on most imported ethanol.[12] A recent survey of federal and state government subsidies in support of ethanol production reported that total annual federal support fell somewhere in the range of $5.4 to $6.6 billion per year—nearly $1 per gallon.[13]

The new RFS in EISA will increase these subsidies dramatically during the life of the program. Based on CRS calculations, federal biofuels subsidies will exceed $25 billion in 2022. Total liability from 2008 through 2022 is estimated at $200 billion.

Table 3. U.S. Farm Prices for Major Agricultural Commodities

Commodity	Unit	Farm Market Prices				USDA Program Prices[a]	
		Average 1997-2006	Actual 2006/2007	Projections		Loan Rate	Target Price
				2007/2008[b]	2012/2013[c]		
Wheat[d]	$/bu	3.24	4.26	6.48	4.29	2.75	3.92
Corn[d]	$/bu	2.15	3.04	4.25	3.25	1.95	2.63
Sorghum[d]	$/bu	2.04	3.29	4.15	3.02[e]	1.95	2.57
Barley[d]	$/bu	2.38	2.85	4.02	3.11[e]	1.85	2.44
Oats[d]	$/bu	1.54	1.87	2.63	1.90[e]	1.33	1.44
Rice[d]	$/cwt	7.17	9.96	12.60	9.64	6.50	10.50
Soybeans[d]	$/bu	5.72	6.43	10.15	7.72	5.00	5.80
Soybean oil[f]	¢/lb	21.4	31.0	53.0	36.8	—	—
Soybean meal[f]	$/st	187.7	205.4	335.0	202.0	—	—
Cotton, Upland	¢/lb	50.3	46.5	57.0[e]	59.9	52.0	72.4
Choice Steers[g]	$/cwt	73.5	85.4	91.8	86.4	—	—
Barrows/Gilts[g]	$/cwt	42.2	47.3	47.1	54.3	—	—
Broilers[g]	¢/lb	37.9	64.4	76.4	77.2	—	—
Eggs[g]	¢/doz	63.7	71.8	127.7	85.4[e]	—	—
Milk[g]	$/cwt	13.91	12.90	19.13	15.7	—	—

Source: Prepared by CRS using data from sources below.

a. For more information on U.S. commodity programs see CRS Report RL34594, *Farm Commodity Programs in the 2008 Farm Bill*, by Jim Monke.
b. Unless otherwise indicated: midpoint of price projection range from USDA, *World Agricultural Supply and Demand Estimates* (WASDE), November 10, 2008.
c. Unless otherwise indicated: FAPRI, *Baseline Update for U.S. Agricultural Markets*, August 2008.
d. Season average farm price from USDA, National Agricultural Statistical Service, *Agricultural Prices.*—= no loan rate.
e. FAPRI, *U.S. Baseline Briefing Book, March 2008,* FAPRI-UMC Report #03-08.
f. USDA, Agr. Marketing Service (AMS), Decatur, IL, cash price, simple average crude for soybean oil, and simple average 48% protein for soybean meal.
g. Calendar year data for the first year, e.g., 2000/2001 = 2000; USDA, AMS: choice steers—Nebraska, direct 1100-1300 lbs.; barrows/gilts—national base, live equivalent 51%-52% lean; broilers—wholesale, 12-city average; eggs—Grade A, New York, volume buyers; and milk—simple average of prices received by farmers for all milk.

Market participants, economists, and biofuels skeptics have begun to question the need for continued large federal incentives in support of ethanol production, particularly when the sector would have been profitable during much of 2006 and 2007 without such subsidies;[14] currently, profitability is less certain, and varies from company to company depending on the amount of debt carried by each company. In addition to opportunity costs, their concerns focus on the potential for widespread unintended consequences that might result from excessive federal incentives adding to the rapid expansion of ethanol production capacity and the demand for corn to feed future ethanol production. These questions extend to issues concerning the ability of the gasoline-marketing infrastructure and auto fleet to accommodate higher ethanol concentrations in gasoline, the likelihood of modifications in engine design, environmental impacts of increased ethanol production and use, and other considerations.

Overview of Non-Corn-Starch-Ethanol RFS Issues

Although most references to "advanced biofuels" involve cellulosic ethanol, much of the "advanced biofuels" component of the EISA RFS may be met by essentially any non-corn-starchderived biofuels. News reports often refer to cellulosic ethanol as "nearing a break-through" or "just around the corner," but the reality is that there is considerable uncertainty about the speed with which this technology may become commercially viable even with substantial government support. A major barrier to cellulosic fuel production is that production costs remain significantly higher than for corn ethanol or other alternative fuels. Many scientists still suggest that commercialization of cellulosic ethanol is 5 to 15 years down the road.[15] Although research is ongoing, presently no commercial-scale cellulosic biofuel plants exist in the United States, and there are only a few demonstration-scale plants in the United States and Canada.[16] Currently, various production processes are prohibitively expensive, including physical, chemical, enzymatic, and microbial treatment and conversion of these feedstocks into motor fuel. For more information on cellulosic biofuels, please see CRS Report RL34738, *Cellulosic Biofuels: Analysis of Policy Issues for Congress*, by Tom Capehart.

Unintended Policy Outcomes of the "Advanced Biofuels" Mandate

Because the advanced biofuel mandate in the RFS is a fixed mandate, irrespective of prices, the above uncertainties about the production of cellulosic ethanol could have significant implications for fuel supply and fuel prices. If cellulosic ethanol production is unable to advance rapidly enough to meet the RFS mandate for non-corn-starch ethanol, then other unexpected biofuels sources may step in and fill the void, such as:

- domestic sorghum-starch ethanol, production of which may expand across the prairie states and in other regions less suitable for corn production;
- domestic sugar-beet ethanol or even costlier domestic biodiesel production may be undertaken to fill the mandate, and could be costly; or
- imports of Brazilian sugar-cane ethanol could expand.

Potential Benefits of the Mandate

Ethanol and biodiesel produced from cellulosic feedstocks, such as prairie grasses and fast- growing trees or agricultural waste, have the potential to improve the energy and environmental effects of U.S. biofuels while offering significant cost savings on the production side (e.g., high- yielding, grown on marginal land, perennial rather than annual). Further, moving away from feed and food crops to dedicated energy crops could avoid some of the agricultural supply and price concerns associated with corn ethanol (as discussed later in this report).

A key potential benefit of many cellulosic feedstocks is that many can be grown without chemicals. Reducing or eliminating chemical fertilizers would address one of the largest energy inputs for corn-based ethanol production. Using biomass to power a biofuels production plant could further reduce fossil fuel inputs. Improving the net energy balance of ethanol would also reduce net fuel-cycle greenhouse gas emissions, although land use change has also been raised as a potential cause of increased greenhouse gas emissions, depending on the type of land used for the feedstock production.

Cellulosic Biofuels Production Uncertainties

There are substantial uncertainties regarding both the costs of producing cellulosic feedstock as well as the costs of producing biofuels from them. Perennial crops are often slow to establish and can take several years before a marketable crop is produced. Crops heavy in cellulose tend to be bulky and represent significant problems in terms of harvesting, transporting, and storing. Seasonality issues involving the operation of a biofuels plant year-round based on a four- or five- month harvest period of biomass suggest that bulkiness is likely to matter a great deal. In addition, most marginal lands (i.e., the low-cost biomass production zones) are located far from major urban markets, making it difficult to reconcile plant location with the cost of fuel distribution.

Further, increases in per-acre yields would be required to make most cellulosic energy crops for fuel production economically competitive. Questions remain whether high yields can be achieved without the use of fertilizers and pesticides. Another question is whether there is sufficient feedstock supply available. USDA estimates that, by 2030, 1.3 billion tons of biomass could be available annually for bioenergy production (including electricity from biomass, and fuels from corn and cellulose).[17] From that, enough biofuels could be produced to replace roughly 70 billion gallons of gasoline per year (about 4.5 million barrels per day). However, this projection assumes technological breakthroughs and significant increases in per-acre yields and, according to USDA, should be seen as an upper bound on what is possible. Further, new harvesting machinery would need to be developed to guarantee an economic supply of cellulosic feedstocks.[18]

In addition to the above concerns, other potential environmental drawbacks associated with cellulosic fuels will need to be addressed, such as the potential for soil erosion, runoff, and the spread of invasive species (many potential biofuels crops are invasive species when introduced into non-native localities). In the near term, the obvious choice of using corn stover[19] to fuel existing corn ethanol plants has its own set of environmental trade-offs, paramount of which is the dilemma of sacrificing soil fertility gains from no- or minimum-tillage corn production.

Energy Supply Issues

Biofuels are not primary energy sources. Energy stored in biological material (through photosynthesis) must be converted into a more useful, portable fuel. This conversion requires energy. The amount and types of energy used to produce biofuels, and the feedstocks for biofuels production, are of key concern. Because of the input energy requirements, the energy and environmental benefits of biofuels and corn ethanol, particularly, may be limited.

Energy Balance

A frequent argument for the use of ethanol as a motor fuel is that it reduces U.S. reliance on oil imports, making the U.S. less vulnerable to disruptions of U.S. oil imports. However, while use of corn ethanol as an alternative fuel displaces petroleum, its overall effect on total energy consumption is less clear. To analyze the net energy consumption of ethanol, the entire fuel cycle must be considered. The fuel cycle consists of all inputs and processes involved in the development, delivery and final use of the fuel. For corn-based ethanol, these inputs include the energy needed to produce fertilizers, operate farm equipment, transport corn, convert corn to ethanol, and distribute the final product. Some studies find a significant positive energy balance of 1.5 or greater—in other words, the energy contained in a gallon of corn ethanol is 50% higher than the amount of energy needed to produce and distribute it. However, other studies suggest that the amount of energy needed to produce ethanol is greater than the amount of energy obtained from its combustion. A review of research studies on ethanol s energy balance and greenhouse gas emissions found that most studies give corn-based ethanol a slight positive energy balance of about 1.2.[20]

If, instead, cellulosic biomass or other feedstocks were used to produce biofuels, the energy balance could be improved. It is expected that most biofuels feedstocks other than corn in the future will require far less nitrogen fertilizer (produced from natural gas) when grown at large scale. Further, if biomass were used to provide process energy at the biofuels refinery, then the energy balance could be even greater. Some estimates are that cellulosic ethanol could have an energy balance of 8.0 or more.[21] Similarly high energy balances have been calculated for sugarcane ethanol and biodiesel.

An expanded RFS would certainly displace petroleum consumption, but the overall effect on lifecycle fossil fuel consumption is questionable, especially if there is a large reliance on corn-based ethanol. EISA requires an increasing amount of "advanced biofuels" resulting in reduced fossil fuel consumption relative to gasoline on a per-mile basis. As the share of advanced biofuels grows, this effect accelerates. However, by 2022, advanced biofuels will likely represent less than 10% of gasoline energy demand, so the total amount of fossil energy displaced would be less than the expected growth in fossil energy consumption from passenger transportation over the same time period.[22]

Natural Gas Demand

As ethanol production increases, the energy needed to process the corn into ethanol, which is produced primarily using natural gas in the United States, can be expected to increase. For example, if the entire 6.5 billion gallons of ethanol produced in 2007 used natural gas as a processing fuel, it would have required an estimated 315 to 380 billion cubic feet (cu. ft.) of natural gas.[23] If the entire 2007 corn crop of 13.1 billion bushels were converted into ethanol, the energy requirements would be equivalent to approximately 1.8 to 2.1 trillion cu. ft. of natural gas. This would have represented about 8% to 9% of total U.S. natural gas consumption, which was an estimated 23.1 trillion cu. ft. in 2007.[24] The United States has been a net importer of natural gas since the early 1980s. A significant increase in its use as a processing fuel in the production of ethanol—and a feedstock for fertilizer production—would likely increase U.S. demand for natural gas.

The EISA RFS proposal boosts corn ethanol production to 15 billion gallons by 2015, requiring an increase in natural gas and/or fertilizer consumption. After 2015, annual eligible corn-starch ethanol under the RFS is capped at 15 billion gallons and advanced biofuels account for increases in renewable fuel use. At that point, demand for natural gas in the biofuels sector will likely stabilize along with ethanol production.

Energy Security[25]

Despite the fact that ethanol displaces gasoline, the benefits to energy security from ethanol are not certain. As stated above, while roughly 31% of the U.S. corn crop is used for ethanol, ethanol only accounts for approximately 3% of gasoline consumption on an energy equivalent basis.[26] The import share of U.S. petroleum consumption was estimated at 60% in 2007, and is expected to grow to 70% by 2025.[27] Further, as long as ethanol remains dependent on U.S. agricultural supplies, any threats to these supplies (such as drought), or increases in crop prices, would negatively affect the feedstock supply and raise the cost of producing enough biofuels to meet the mandate. In fact, in 1995 high corn prices—due to strong export demand—contributed to an 18% decline in ethanol production between 1995 and 1996.

Moreover, expanding corn-based ethanol production to levels needed to significantly promote U.S. energy security is likely to be infeasible. If the entire 2007 U.S. corn crop of 13.1 billion bushels were used as ethanol feedstock, the resultant 37 billion gallons of ethanol (24.6 billion gasoline-equivalent gallons (GEG)) would represent about 17% of estimated national gasoline use of approximately 143 billion gallons.[28] In 2008, a projected 78.2 million acres of corn were harvested (second largest since 1944). Nearly 137 million acres would be needed to produce enough corn (20.5 billion bushels) and resulting ethanol (56.4 billion gallons or 37.8 billion GEG) to substitute for roughly 20% of petroleum imports.[29] Thus, barring a drastic realignment of U.S. field crop production patterns, corn-based ethanol's potential as a petroleum import substitute appears to be limited by crop area constraints, among other factors.[30]

The specific definition of "advanced biofuels" also affects the overall energy security picture for biofuels. For example, if ethanol from sugarcane is imported under an expanded RFS; this provides an incentive to increase imports of sugarcane ethanol, especially from Brazil. The expanded RFS also provides an incentive for imports of biodiesel and other renewable diesel substitutes from tropical countries.

Energy Prices

The effects of the expanded RFS on energy prices are uncertain. If wholesale biofuels prices remain higher than gasoline prices (after all economic incentives are taken into account), then mandating higher and higher levels of biofuels would likely lead to higher gasoline pump prices. However, if petroleum prices—and thus gasoline prices—are high, the use of some biofuels might help to mitigate high gasoline prices.

Current production costs are so high for some biofuels, especially cellulosic biofuels and biodiesel from algae, that significant technological advances—or significant increases in petroleum prices—are necessary to lower their production costs to make them competitive with gasoline. Without cost reductions, mandating large amounts of these fuels would likely raise fuel prices. If a price were placed on greenhouse gas emissions—perhaps through the enactment of a cap and trade bill—then the economics could shift in favor of these fuels despite their high production costs, as they have lower fuel-cycle and life-cycle greenhouse gas emissions (see below).

Greenhouse Gas Emissions

Biofuels proponents argue that a key benefit of biofuels use is a decrease in greenhouse gas (GHG) emissions. However, some question the overall GHG benefit of biofuels, especially corn-based ethanol. There is a wide range of fuel-cycle estimates for greenhouse gas reductions from corn-based ethanol. However, most studies have found that corn-based ethanol reduces fuel-cycle GHG emissions by 10%-20% per mile relative to gasoline.[31] These estimates vary depending on several factors including the cultivation practice (e.g., minimum-tillage versus normal tillage) used to grow the corn and the fuel used to process the corn into ethanol (e.g., natural gas versus coal). These same studies find that biofuels produced from sugar cane or cellulosic biomass could reduce fuel-cycle GHG emissions by as much as 90% per mile relative to gasoline.

However, fuel-cycle analyses generally do not take changes in land use into account. For example, if a previously uncultivated piece of land is tilled to plant biofuels crops, some of the carbon stored in the field could be released. In that case, the overall GHG benefit of biofuels could be compromised.[32] One study estimates that taking land use into account (a life-cycle analysis, as opposed to a fuel-cycle analysis), the GHG reduction from corn ethanol is less than 3% per mile relative to gasoline,[33] while cellulosic biofuels have a life-cycle reduction of 50%.[34] Other recent studies indicate even smaller GHG reductions.

Biofuels produced at facilities commencing operations after the date of enactment must have a 20% life-cycle emissions reduction to qualify under the EISA expanded RFS. However, it is expected that this provision may not be relevant to a large share of conventional ethanol since much of the capacity to meet the 15 billion gallon cap currently exists or will come from expansions of existing plants.

Agricultural Issues

A continued expansion of corn-based ethanol production could have significant consequences for traditional U.S. agricultural crop production and rural economies. Supporters of an expanded RFS claim that increased biofuels production and use would have enormous agricultural and rural economic benefits by increasing farm and rural incomes and generating substantial rural employment opportunities.[35] However, large-scale shifts in agricultural production activities will likely also have important regional economic consequences that have yet to be fully considered or understood. As corn prices rise, so too does the incentive to expand corn production either by expanding production to more marginal soil environments or by altering the traditional corn- soybean rotation that characterizes Corn Belt agriculture. This shift could displace other field crops, primarily soybeans, and other agricultural activities. Further, corn production is among the most energy-intensive of the major field crops. An expansion of corn area would likely have important and unwanted environmental consequences due to the increases in fertilizer and chemical use and soil erosion. The National Corn Growers Association estimates that U.S. corn-based ethanol production could expand to between 12.8 and 17.8 billion gallons by 2015 without significantly affecting agricultural markets.[36] However, as noted below, other evidence suggests effects, such as higher commodity prices, are already being felt in the current expansion in corn production.

Food versus Fuel

Many critics of federal biofuels subsidies and the RFS argue that a sustained rise in grain prices driven by ethanol feedstock demand likely will lead to higher U.S. and world food prices with potentially harmful effects on consumer budgets and nutrition.[37] As evidence they cite USDA s estimate that the U.S. Consumer Price Index (CPI) for all food was forecast to increase 5%-6% in 2008, and increased 4.0% in 2007, and 2.4% in 2006.[38] The average rate of increase for 1997- 2006 was 2.5%.[39] Lower fuel and commodity prices are forecast to lower the increase in the CPI for all food to 3.5% to 4.5% in 2009.[40] However, in analyzing this argument it is important to distinguish between prices of farm-level crops and retail-level food products because most "food" prices are largely determined by costs and profits after the commodities leave the farm.[41] Basic economics suggests that the price of a particular retail food item varies with a change in the price of an underlying ingredient in direct relation to the relative importance (in value terms) of that ingredient. For example, if the value of wheat in a $1.00 loaf of bread is about 10 cents, then a 20% rise in the price of wheat translates into a 2-cent rise in a loaf of bread.

As a result of corn's relatively small value-share in most retail food product prices, it is unlikely that the ethanol-driven corn price surge is a major factor in current food price inflation estimates.[42] Furthermore, economists generally agree that most retail food price increases are not due to ethanol-driven demand increases, but rather are the result of two major factors—a sharp increase in energy prices which ripples through all phases of marketing and processing channels, and the strong increase in demand for agricultural products in the international marketplace from China and India (a product of their large populations and rapid economic growth).[43]

Feed Markets

Most corn grown in the United States is used for animal feed. From 1995 through 2005, domestic feed use accounted for 58% of U.S. corn use. As corn-based ethanol production increases, so do total corn demand and corn prices. As a result, sustained higher corn prices likely will have significant consequences for traditional feed markets and the livestock industries—hog, cattle, dairy, and poultry—that depend on those feed markets. Corn traditionally has represented about 57% of feed concentrates and processed feedstuffs fed to animals in the United States.[44] Persistent high feed costs will tighten profit margins and likely squeeze out marginal livestock producers. Because economies of scale tend to favor larger producers, persistently tighter profit margins suggest a potential for increased concentration in the livestock sector. The National Cattlemen s Beef Association (NCBA) has been one of the foremost critics of an expanded RFS. Instead, the NCBA argues for a phase out of current ethanol subsidies and a more market-based approach to renewable fuels policy.[45]

The price of corn also is linked to the price of other grains, including those destined for food markets, through competition in the feed marketplace and in the producers planting choices for limited acreage. The price runup in the U.S. corn market has already spilled over into price increases in the markets for soybeans and soybean oil. Supply distortions also are likely to develop in protein-meal markets related to expanded production of the ethanol processing byproduct, distillers dried grains with solubles (DDGS), which averages about 30% protein content and can substitute in certain feed and meal markets.[46] Although DDGS use would substitute for some of the lost feed value of corn used in ethanol processing, about 66% of the original weight of corn is consumed in producing ethanol and is no longer available for feed. Furthermore, not all livestock species are well adapted to dramatically increased consumption of DDGS in their rations—dairy cattle appear to be best suited to expanding DDGS s share in feed rations; poultry and pork are much less able to adapt. Also, DDGS must be dried before it can be transported long distances, adding to feed costs and consuming more fuel. There may be some potential for large- scale livestock producers to relocate near new feed sources, but such relocation likely would have important regional economic effects.

Domestic Food Prices

Although corn primarily is used as a livestock feed or for ethanol production, it is also used widely as an ingredient (albeit minor) in many processed foods, for example, soft drinks, snack foods, and baked goods. Since corn prices are a relatively small share of the price of most retail food products, their price impact is concomitantly small. Higher corn prices have their largest impact on meat prices. The feed-price effect will first translate into higher prices for poultry and hogs, which are less able to use alternate feedstuffs. Dairy and beef cattle are more versatile in their ability to shift to alternate feed sources, but eventually a sustained rise in corn prices will push their feed costs upward as well. A recent economic study estimated that a 30% increase in the price of corn, and associated increases in the prices of wheat and soybeans, would increase egg prices by 8.1%, poultry prices by 5.1%, pork prices by 4.5%, beef prices by 4.1%, and milk prices by 2.7%.[47] The effect was a 1.1% increase (0.9% on at-home food and 1.3% on awayfrom-home food consumption) in the all-food CPI. Thus, the price impact of higher corn prices is small but important for most livestock products, and probably much smaller for most other retail food products.

The overall impact to consumers from higher food prices depends on the proportion of income that is spent on food. Since food costs represent a relatively small share of consumer spending for most U.S. households (about 10%), food price increases (from whatever source) are absorbed relatively easily in the short run. However, low-income consumers spend a much greater proportion of their income on food than do high-income consumers. Their larger share combined with less flexibility to adjust expenditures in other budget areas means that any increase in food prices potentially could cause hardship.[48] In addition, higher commodity prices combined with shrinking inventories mean that local school districts and the U.S. government will be forced to pay higher market prices for food for school lunch programs. The automatic food price escalators built into the food stamp program, renamed as Supplemental Nutrition Assistance Program (SNAP), mean rising expenditures as well.[49]

International Food Prices

Due to trade linkages, the increase in U.S. corn prices has become a concern for international markets as well. High commodity prices ripple through international markets where impacts vary widely based on grain import dependence and the ability to respond to higher commodity prices. Import-dependent developing country markets are put at greater food security risk due to the higher cost of imported commodities. In particular, lower-income households in many foreign markets where food imports are an important share of national consumption and where food expenses represent a larger portion of the household budget may be affected by higher food prices.[50] In China, where corn is an important food source, the government recently has put a halt to its planned ethanol plant expansion due to the threat it poses to the country s food security. Similarly, humanitarian groups have expressed concern for the potential difficulties that higher grain prices imply for developing countries that are net food importers.[51]

Exports

The United States is the worlds leading exporter of corn. In the past decade (1997 to 2006), the United States has exported about 20% of its corn production, accounting for nearly 66% of world corn trade.[52] Increased use of corn for ethanol production could reduce the volume of U.S. corn production available for export. In 2006, the volume of corn used for ethanol equaled exports, with a 20% share of total use. By the 2009/10 marketing year (September-August), ethanol s share of U.S. corn production is expected to reach nearly 36%, while the export share falls to 14%.[53] FAPRI projections clearly suggest that higher corn prices will result in lost export sales. It is unclear what type of market adjustments will occur in global feed markets, since several different grains and feedstuffs are relatively close substitutes. Price-sensitive corn importers may quickly switch to alternative, cheaper sources of feed, depending on the availability of supplies and the adaptability of animal rations. In contrast, less price-sensitive corn importers, such as Japan and Taiwan, may choose to pay a higher price in an attempt to bid the corn away from ethanol plants. There could be significant economic effects to U.S. grain companies and to the U.S. agricultural sector if ethanol-induced higher corn prices led to a sustained reshaping of international grain trade.

Economic Impact

Several studies claim that increased biofuels production and use would produce enormous agricultural and rural economic benefits by raising farm and rural incomes and generating substantial rural employment opportunities.[54] One estimate suggested that the economic benefit from the ethanol industry to the U.S. economy for 2005 was $17.7 billion of GDP; the creation of over 150,000 jobs; $5.7 billion in spinoff economic activity; and more than $3.5 billion in government tax revenues.[55] However, a recent critical review of the standard input-output methodology used to generate such economic impact estimates suggests that the income and job growth attributable to biofuels production has been grossly overstated, perhaps by as much as a factor of four or five.[56] Yet, while the magnitude may be called into question, there appears to be no doubt about the potential positive value of biofuels production to rural economies. First, in addition to temporary construction work to build a new plant, several dozen permanent jobs also accompany a biofuels plant depending on the plants operating capacity. Second, the new demand boosts the local prices received by farmers for corn and sorghum. Third, important secondary economic activity is associated with the operation of an ethanol plant. Fourth, given the high level of federal and state subsidies for the biofuels industry, any locality that is home to a biofuels plant can expect substantial net transfers of government funds into the areas economy.

The policy question of interest is not whether there are positive gains from growth in the ethanol industry, but whether the growth and its economic implications are sufficient to merit large government subsidies. A growing number of critics argue that the answer is no.[57] Others suggest that, at the very least, the issue deserves more study before continuing or expanding current government support levels.[58]

Ethanol Infrastructure and Distribution Issues

In addition to the above concerns about raw material supply for ethanol production (both feedstock and energy), there are issues involving ethanol distribution and infrastructure. Expanding ethanol production likely will strain the existing supply infrastructure. Further, expansion of ethanol use beyond the current 10% blend will require investment in entirely new infrastructure that would be necessary to handle a higher and higher percentage of ethanol in gasoline. If biomass-based diesel substitutes are produced in much larger quantities, some of these infrastructure issues may be mitigated.

Distribution Issues

Ethanol-blended gasoline tends to separate in pipelines due to the presence of water in the lines. Further, ethanol is corrosive and may damage existing pipelines and storage tanks. Therefore, unlike petroleum products, ethanol and ethanol blended gasoline cannot be shipped by pipeline in the United States. Another issue with pipeline transportation is that corn ethanol must be moved from rural areas in the Midwest to more populated areas, which are often located along the coasts. This shipment is in the opposite direction of existing pipeline transportation, which moves gasoline from refiners along the coast to other coastal cities and into the interior of the country. While some studies have concluded that shipping ethanol or ethanol-blended gasoline via pipeline could be feasible, no major U.S. pipeline has made the investments to allow such shipments.[59]

Thus, the current distribution system for ethanol is dependent on rail cars, tanker trucks, and barges. These deliver ethanol to fuel terminals where it is blended with gasoline before shipment via tanker truck to gasoline retailers. However, these transport modes lead to prices higher than for pipeline transport, and the supply of current shipping options (especially rail cars) is limited. For example, according to industry estimates, the number of ethanol carloads has tripled between 2001 and 2006, and the number is expected to increase by another 30% in 2007, although final data is not yet available.[60] A significant increase in corn-based ethanol production would further strain this tight transport situation.

Because of these distribution issues, some pipeline operators are seeking ways to make their systems compatible with ethanol or ethanol-blended gasoline. These modifications could include coating the interior of pipelines with epoxy or some other, corrosion-resistant material. Another potential strategy could be to replace all susceptible pipeline components with newer, hardier components. However, even if such modifications are technically possible, they likely will be expensive, and could further increase ethanol transportation costs.

As non-corn biofuels play a larger role, as required in EISA, some of the supply infrastructure concerns may be alleviated. Cellulosic biofuels potentially can be produced from a variety of feedstocks, and may not be as dependent on a single crop from one region of the country. For example, municipal solid waste is ubiquitous across the United States, and could serve as a ready feedstock for biofuels production if the technology were developed to convert it economically to fuel. Further, increased imports of biofuels from other countries could allow for greater use of biofuels, especially along the coasts. Moreover, some biofuels, especially some diesel substitutes, may be able to be mixed with petroleum fuels at the refinery and placed directly into the pipeline.

Higher-Level Ethanol Blends

One key benefit of gasoline-ethanol blends up to 10% ethanol is that they are compatible with existing vehicles and infrastructure (e.g., fuel tanks, retail pumps, etc.). All automakers that produce cars and light trucks for the U.S. market warranty their vehicles to run on gasoline with up to 10% ethanol (E10). This 10% currently is an upper bound (sometimes referred to as the "blend wall") to the amount of ethanol that can be introduced into the gasoline pool. If most or all gasoline in the country contained 10% ethanol, this would allow only for roughly 15 billion gallons, far less than the amount of biofuels mandated in EISA.

As a major producer of ethanol for its domestic market, Brazil has a mandate that all of its gasoline contain 20-25% ethanol. For the United States to move to E20 (20% ethanol, 80% gasoline), it may be that few (if any) modifications would need to be made to existing vehicles and infrastructure. Vehicle testing, however, would be necessary to determine whether new vehicle parts would be required, or if existing vehicles are compatible with E20. Similar testing would be necessary for terminal tanks, tanker trucks, retail storage tanks, pumps, etc. In addition, EPA would need to certify that the fuel will not lead to increased air quality problems.

There is also interest in expanding the use of E85 (85% ethanol, 15% gasoline). Current E85 consumption represents only about 1% of ethanol consumption in the United States. A key reason for the relatively low consumption of E85 is that relatively few vehicles operate on E85. The National Ethanol Vehicle Coalition estimates that there are approximately six million E85- capable vehicles on U.S. roads,[61] as compared to approximately 230 million gasoline- and diesel- fueled vehicles.[62] Most E85-capable vehicles are flexible fuel vehicles or FFVs. An FFV can operate on any mixture of gasoline and between 0% and 85% ethanol. However, ethanol has a lower per gallon energy content than gasoline. Therefore, FFVs tend to have lower fuel economy when operating on E85. For the use of E85 to be economical, the pump price for E85 must be low enough to make up for the decreased fuel economy relative to gasoline. Generally, to have equivalent per-mile costs, E85 must cost 20% to 30% less per gallon at the pump than gasoline. Owners of a large majority of the FFVs on U.S. roads choose to fuel them exclusively with gasoline, largely due to higher per-mile fuel cost and lower availability of E85.

E85 capacity is expanding rapidly, with the number of E85 stations nearly tripling between January 2006 and January 2008. But those stations still represent less than 1% of U.S. gasoline retailers. Further expansion will require significant investments, especially at the retail level. Installation of a new E85 pump and underground tank can cost as much as $100,000 to $200,000.[63] However, if existing equipment can be used with little modification, the cost could be less than $10,000.

Vehicle Infrastructure Issues

As was stated above, if a large portion of any increased RFS is met using ethanol, then the United States likely does not have the vehicles to consume the fuel. The 10% blend wall on ethanol in gasoline for conventional vehicles poses a significant barrier to expanding ethanol consumption beyond 15 billion gallons per year.[64] To allow more ethanol use, vehicles will need to be certified and warranted for higher-level ethanol blends, or the number of ethanol FFVs will need to increase. Turnover of the U.S. automobile fleet is likely to slow during the recession, making it more difficult to integrate FFVs into the fleet.

CONCLUSION

There is continuing interest in expanding the U.S. biofuels industry as a strategy for promoting energy security and achieving environmental goals. However, it is possible that increased biofuel production may place desired policy objectives in conflict with one another. There are limits to the amount of biofuels that can be produced from current feedstocks and questions about the net energy and environmental benefits they might provide. Further, rapid expansion of biofuels production may have many unintended and undesirable consequences for agricultural commodity costs, fossil energy use, and environmental degradation. Owing to these concerns, alternative strategies for energy conservation and alternative energy production are widely seen as warranting consideration.

End Notes

[1] The term advanced biofuels comes from legislation in the 110th Congress, and is defined in Section 201 of the Energy Independence and Security Act of 2007 (EISA). In many cases, the definition of advanced biofuels includes mature technologies and fuels that are currently produced in large amounts. For example, the EISA definition of advanced biofuels potentially includes ethanol from sugarcane, despite the fact that Brazilian sugar growers have been producing fuel ethanol for decades. EISA defines advanced biofuels as biofuels other than ethanol derived from corn starch (kernels) having 50% lower lifecycle greenhouse gas emissions relative to gasoline. Possible fuels include biodiesel from oil seeds, ethanol from sugarcane, and ethanol from cellulosic materials (including non-starch parts of the corn plant, such as the stalk).

[2] An additional roughly 4 billion gallons were imported, mostly from Brazil and Caribbean Basin Initiative (CBI) countries.

[3] In 2006 ethanol prices rose sharply, and direct imports from Brazil rose sharply, despite the tariff.

[4] For more information on CBI imports, see CRS Report RS21930, *Ethanol Imports and the Caribbean Basin Initiative (CBI)*, by Brent D. Yacobucci.

[5] DOE-EIA Annual Energy Review 2007, Report No. DOE/EIA-0384(2007), June 23, 2008, http://www.eia.doe.gov/emeu/aer/pdf/pages/sec10_11.pdf.

[6] U.S. Energy Information Administration, *U.S. Product Supplied for Crude Oil and Petroleum Products*, Washington, DC, July 28, 2008, http://tonto.eia.doe.gov/dnav/pet/pet_cons_psup_dc_nus_mbbl_a.htm.

[7] For example, see Bruce A. Babcock, High Crop Prices, Ethanol Mandates, and the Public Good: Do They Coexist? *Iowa Ag Review*, Vol. 13, No. 2, Spring 2007; and Robert Hahn

and Caroline Cecot, The Benefits and Costs of Ethanol, Working Paper 07-17, AEI-Brookings Joint Center for Regulatory Studies, November 2007.
[8] USDA, WAOB, *World Agricultural Supply and Demand Estimates (WASDE) Report*, December 11, 2008, Washington; available at [http://www.usda.gov/oce/].
[9] See Renewable Fuels Association, *Industry Statistics*, at http://www.ethanolrfa.org/industry
[10] FAPRI, *Baseline Update for U.S. Agricultural Markets*, FAPRI-MU report #28-07, August 2007.
[11] No. 2 yellow, Central Illinois; USDA Agricultural Marketing Service; Ethanol are rack, f.o.b. Omaha, Nebraska Ethanol Board, Lincoln, NE., Nebraska Energy Office, Lincoln, NE.
[12] The blender's tax credit declines to $0.45 per gallon the first year following that in which annual production and imports exceed 7.5 billion gallons. This level is expected to have been reached in 2008 making the reduction effective in 2009. For more information on incentives (both tax and non-tax) for ethanol, see CRS Report RL33 572, *Biofuels Incentives: A Summary of Federal Programs*, by Brent D. Yacobucci.
[13] Ronald Steenblik. *Biofuels–At What Cost? Government Support for Ethanol and Biodiesel in the United States*, Global Subsidies Initiative of the International Institute for Sustainable Development, Geneva, Switzerland, September 2007, p. 37; available at http://www.globalsubsidies.org.
[14] Chris Hurt, Wally Tyner, and Otto Doering, Department of Agricultural Economics, Purdue University, *Economics of Ethanol*, December 2006, West Lafayette, IN.
[15] For example, the Department of Energy s goal is to make cellulosic biofuels cost-competitive with corn ethanol by 2012. Other groups are less optimistic.
[16] However, on February 28, 2007, DOE announced availability of $385 million in grant funding for six commercial- scale cellulosic ethanol plants in six states. If operational, combined capacity of these six plants would be 130 million gallons per year. DOE, *DOE Selects Six Cellulosic Ethanol Plants for Up to $385 Million in Federal Funding*, February 28, 2007, Washington, D.C. Subsequently, 2 projects were cancelled by the recipients.
[17] Oak Ridge National Laboratory for DOE and USDA, *Biomass as a Feedstock for a Bioenergy and Bioproducts Industry: The Technical Feasibility of a Billion-Ton Annual Supply*, April 2005, Oak Ridge, TN.
[18] For example, the study assumes roughly 400 million tons of biomass from agricultural residues. To economically supply those residues to biofuels producers, farm equipment manufacturers likely would need to develop one-pass harvesters that could collect and separate crops and crop residues at the same time.
[19] Stover is the above-soil part of the corn plant excluding the kernels.
[20] Alexander E. Farrell, Richard J. Plevin, Brian T. Turner, Andrew D. Jones, Michael O Hare, and Daniel M. Kammen, Ethanol Can Contribute to Energy and Environmental Goals, *Science*, Jan. 27, 2006, pp. 506-508.
[21] David Andress, *Ethanol Energy Balances*. November 2002.
[22] For example, EIA projects that motor gasoline consumption will increase 22% between 2007 and 2011. EIA, *Annual Energy Outlook*. Table 11.
[23] CRS calculations based on energy usage rates of 49,733 Btu/gal of ethanol from Shapouri (2004), roughly 60,000 Btu/gal from Farrell (2006). Hosein Shapouri and Andrew McAloon, USDA, Office of the Chief Economist, *The 2001 Net Energy Balance of Corn-Ethanol*, 2004, Washington; Farrell, op. cit.
[24] U.S. Department of Energy (DOE), Energy Information Administration (EIA), *Annual Energy Outlook 2007 with Projections to 2030*, Table 1, Total Energy Supply and Disposition Summary, Washington; at http://www.eia.doe.gov/ oiaf/aeo/index.html.
[25] A key question in evaluating the energy security benefits or costs of an expanded RFS is what is the definition of energy security. For many policymakers, energy security and energy independence (i.e., producing all energy within our borders) are synonymous. For others,

26 By volume, ethanol accounted for approximately 4.6% of gasoline consumption in the United States in 2006, but a gallon of ethanol yields only 67% of the energy of a gallon of gasoline. energy security means guaranteeing that we have reliable supplies of energy regardless of their origin. For this section, the former definition is used.

27 DOE, EIA, Annual Energy Review 2007, Washington, June 2008, Table 5.1.

28 This estimate is based on USDA s November 10, 2008, *World Agricultural Supply and Demand Estimates (WASDE) Report*, and using comparable conversion rates.

29 This represents roughly half of gasoline s share of imported petroleum. However, petroleum imports are primarily unrefined crude oil, which is then refined into a variety of products. CRS calculations assume corn yields of 150 bushels per acre and an ethanol yield of 2.75 gal/bu.

30 Two recent articles by economists at Iowa State University examine the potential for obtaining a 10 million acre expansion in corn planting: Bruce Babcock and D. A. Hennessy, "Getting More Corn Acres From the Corn Belt"; and Chad E. Hart, "Feeding the Ethanol Boom: Where Will the Corn Come From?" *Iowa Ag Review*, Vol. 12, No. 4, Fall 2006.

31 EPA, Greenhouse Gas Impacts of Expanded Renewable and Alternative Fuels Use, April 2007; Farrell et al.

32 See Timothy Searchinger, Ralph Heimlich, and R. A. Houghton, et al., "Use of U.S. Croplands for Biofuels Increases Greenhouse Gases Through Emissions from Land-Use Change," *Science*, vol. 319, no. 5867 (February 2008).

33 Mark A. Delucchi, Draft Report: Life Cycle Analyses of Biofuels, 2006.

34 While a 50% life-cycle reduction is still significant, it is far less than the 90% reduction suggested some by fuel- cycle analyses.

35 For example, see John M. Urbanchuk (Director, LECG LLC), *Contribution of the Ethanol Industry to the Economy of the United States*, white paper prepared for National Corn Growers Assoc., February 21, 2006.

36 National Corn Growers Association, *How Much Ethanol Can Come From Corn?*, November 9, 2006, Washington, DC.

37 For a discussion, see the National Corn Growers Associations online Food versus Fuel Debate, at http://www.ncga.com/news/OurView/pdf/2006/FoodANDFuel.pdf.

38 USDA Economic Research Service, Food CPI, Prices, and Expenditures Briefing Room, http://www.ers.usda.gov/ briefing/cpifoodandexpenditures/Data/cpiforecasts.htm.

39 ERS, USDA, Briefing Room Food CPI, Prices, and Expenditures, at http://www.ers.usda.gov/Briefing/ CPIFoodAndExpenditures/consumerpriceindex.htm.

40 Ibid.

41 Helen H. Jensen and Bruce A. Babcock, Do Biofuels Mean Inexpensive Food Is a Thing of the Past? *Iowa Ag Review*, Spring 2007, Vol. 13, No. 2, pp. 1-3.

42 For examples, see Food & Water Watch, Retail Realities: Corn Prices Do Not Drive Grocery Inflation, Sept. 2007; and John M. Urbanchuk (Director, LECG LLC), The Relative Impact of Corn and Energy Prices in the Grocery Aisle, white paper prepared for National Corn Growers Association, June, 14, 2007.

43 For examples, see Jacque Diouf, Director General of the U.N. Food and Agriculture Organization, Why Are Food Prices Rising? in *Financial Times Online*, November 26, 2007; http://media.. See also Keith Collins, Chief Economist, USDA, Testimony before the House Committee on Agriculture, October 18, 2007.

44 USDA, ERS, *Feed Situation and Outlook Yearbook*, FDS-2003, April 2003, Washington.

45 NCBA on Renewable Fuel Policy, NCBA Issue Backgrounder-2007; available at http://www.beefusa.org/uDocs/ NCBAonRenewableFuelPolicy-2007.pdf.

46 For a discussion of potential feed market effects due to growing ethanol production, see Bob Kohlmeyer, The Other Side of Ethanol s Bonanza, *Ag Perspectives* (World Perspectives, Inc.), Dec. 14, 2004; and R. Wisner and P. Baumel, Ethanol, Exports, and Livestock: Will There be Enough Corn to Supply Future Needs?, *Feedstuffs*, no. 30, vol. 76, July 26, 2004.

[47] Simla Tokgoz and others, "Emerging Biofuels: Outlook of Effects on U.S. Grain, Oilseed, and Livestock Markets," Staff Report 07-SR 101, Center for Agricultural Research and Development (CARD), Iowa State University, May 2007.

[48] Helen H. Jensen and Bruce A. Babcock, Do Biofuels Mean Inexpensive Food is a Thing of the Past? *Iowa Ag Review*, Spring 2007, Vol. 13, No. 2, pp. 1-3.

[49] Ibid.

[50] Shahla Shapouri and Stacey Rosen, Energy Price Implications for Food Security in Developing Countries, *Food Security Assessment, 2006*, GFA-18, Economic Research Service, USDA.

[51] International Monetary Fund, *World Economic Outlook: Globalization and Inequality*. October 2007. Washington.

[52] USDA, Production, Supply and Distribution Online (PSD database) available at http://www.fas.usda.gov/psdonline/ psdHome.aspx.

[53] FAPRI, Baseline Update for U.S. Agricultural Markets, August 2008.

[54] For example, see John M. Urbanchuk (Director, LECG LLC), *Contribution of the Ethanol Industry to the Economy of the United States*, white paper prepared for National Corn Growers Assoc., February 21, 2006; see also Urbanchuk, *Contribution of the Biofuels Industry To the Economy of Iowa*, white paper prepared for the Iowa Renewable Fuels Association, February 2007.

[55] Urbanchuk (2006).

[56] David Swenson, "Input-Outrageous: The Economic Impacts of Modern Biofuels Production," Department of Economics, Iowa State University (ISU), June 2006. Similar results are found in: David Swenson, "Understanding Biofuels Economic Impact Claims," Department of Economics, ISU, April 2007; Lisa Eathington and Dave Swenson, "Dude, Where s My Corn? Constraints on the Location of Ethanol Production in the Corn Belt," Department of Economics, ISU, paper presented at 46th Annual Meeting of the Southern Regional Science Assoc., Charleston, SC, March 29-31, 2007; Swenson and Eathington, "Determining the Regional Economic Values of Ethanol Production in Iowa Considering Different Levels of Local Investment," Department of Economics, ISU, July 2006.

[57] Examples include Robert Hahn and Caroline Cecot, "The Benefits and Costs of Ethanol," Working Paper 07-17, AEI-Brookings Joint Center for Regulatory Studies, November 2007; Richard Doornbosch and Ronald Steenblik, "Biofuels: Is the Cure Worse Than the Disease?" paper presented at an OECD Round Table on Sustainable Development, Paris, September 11-12, 2007; and Doug Koplow, *Biofuels at What Cost? Government Support for Ethanol and Biodiesel in the United States: 2007 Update*, report prepared for the Global Studies Initiative of the International Institute for Sustainable Development, Geneva, Switzerland, October 2007.

[58] For example, see Bruce A. Babcock, "High Crop Prices, Ethanol Mandates, and the Public Good: Do They Coexist?" *Iowa Ag Review*, vol. 13, no. 2, Spring 2007.

[59] Some small, proprietary ethanol pipelines do exist. American Petroleum Institute, *Shipping Ethanol Through Pipelines*. Available at
http://www.api.org/aboutoilgas/sectors/pipeline/upload

[60] Ilan Brat and Daniel Machalaba, Can Ethanol Get a Ticket to Ride?, *The Wall Street Journal*, Feb. 1, 2007, p. B 1.

[61] National Ethanol Vehicle Coalition, *Frequently Asked Questions*, accessed February 3, 2006, at http://www.e85fuel.com/e85101/faq.php.

[62] Federal Highway Administration, *Highway Statistics 2003*, November 2004, Washington.

[63] David Sedgwick, *Automotive News*, January 29, 2007. p. 112.

[64] Note that 15 billion gallons is the corn starch ethanol limit for the expanded RFS in the EISA.

In: World Biofuels Production Potential
Editor: Thomas E. Rommer

ISBN: 978-1-61668-663-5
© 2010 Nova Science Publishers, Inc.

Chapter 3

WORLD BIOFUELS PRODUCTION POTENTIAL: UNDERSTANDING THE CHALLENGES TO MEETING THE U.S. RENEWABLE FUEL STANDARD

Office of Policy Analysis

ACKNOWLEDGMENTS

This report was prepared under the supervision of Carmen Difiglio, Deputy Assistant Secretary for Policy Analysis, Office of Policy and International Affairs. Primary authors were Bhima Sastri and Audrey Lee in the Office of Policy Analysis. Substantial contributions were also made by Tina Kaarsberg in the same office, Thomas Alfstad of the Brookhaven National Laboratory, Michael Curtis (now at the U.S. Agency for International Development), and Daniel MacNeil (now at PacifiCorp).

This study was funded by the Office of Biomass Programs, Office of Energy Efficiency and Renewable Energy, and is based on analysis provided by the Oak Ridge National Laboratory, the National Renewable Energy Laboratory, and the Brookhaven National Laboratory. The Office of Policy Analysis wants to particularly acknowledge the support and guidance of Jacques BeaudryLosique, Director of the Office of Biomass Programs.

EXECUTIVE SUMMARY

This study by the U.S. Department of Energy (DOE) estimates the worldwide potential to produce biofuels including biofuels for export. It was undertaken to improve our understanding of the potential for imported biofuels to satisfy the requirements of Title II of the 2007 Energy Independence and Security Act (EISA) in the coming decades[1]. Many other countries' biofuels production and policies are expanding as rapidly as ours. Therefore, we modeled a detailed and up-to-date representation of the amount of biofuel feedstocks that are being and can be grown, current and future biofuels production capacity, and other factors relevant to the economic competitiveness of worldwide biofuels production, use, and trade.

The Oak Ridge National Laboratory (ORNL) identified and prepared feedstock data for countries that were likely to be significant exporters of biofuels to the U.S. The National Renewable Energy Laboratory (NREL) calculated conversion costs by conducting material flow analyses and technology assessments on biofuels technologies. Brookhaven National Laboratory (BNL) integrated the country specific feedstock estimates and conversion costs into the global Energy Technology Perspectives (ETP) MARKAL (MARKet ALlocation) model. The model uses least- cost optimization to project the future state of the global energy system in five year increments. World biofuels production was assessed over the 2010 to 2030 timeframe using scenarios covering a range U.S. policies (tax credits, tariffs, and regulations), as well as oil prices, feedstock availability, and a global CO_2 price.

All scenarios include the full implementation of existing U.S. and selected other countries' biofuels' policies (Table 4). For the U.S., the most important policy is the EISA Title II Renewable Fuel Standard (RFS). It progressively increases the required volumes of renewable fuel used in motor vehicles (Appendix B). The RFS requires 36 billion (B) gallons (gal) per year of renewable fuels by 2022[2]. Within the mandate, amounts of advanced biofuels, including biomass-based diesel and cellulosic biofuels, are required beginning in 2009. Imported renewable fuels are also eligible for the RFS. Another key U.S. policy is the $1.01 per gal tax credit for producers of cellulosic biofuels enacted as part of the 2008 Farm Bill[3]. This credit, along with the DOE's research, development and demonstration (RD&D) programs, are assumed to enable the rapid expansion of U.S. and global cellulosic biofuels production needed for the U.S. to approach the 2022 RFS goal[4]. While the Environmental Protection Agency (EPA) has yet to issue RFS rules to determine which fuels would meet the greenhouse gas (GHG) reduction and land use restrictions specified in EISA, we assume that cellulosic ethanol, biomass-to-liquid fuels (BTL), sugar-derived ethanol, and fatty acid methyl ester biodiesel would all meet the EISA advanced biofuel requirements. We also assume that enough U.S. corn ethanol would meet EISA's biofuel requirements or otherwise be grandfathered under EISA to reach 15 B gal per year.

Primary Results from the Scenario Analysis

Many countries around the world are embarking on ambitious biofuel policies through renewable fuel standards and economic incentives. As a result, both global biofuel demand and supply is expected to grow very rapidly over the next two decades, provided policymakers maintain their policy goals. In the reference case presented here, total biofuel production increases more than six-fold from 12 B gal in 2005 to 83 B gal in 2030. The infrastructure challenges are daunting, will require considerable investment, and will test the innovation systems in countries with nascent biofuel industries. The ability to transfer technology and trade in biofuels is essential to meeting new biofuels goals.

Sugar-based ethanol is now the least expensive biofuel and its production is mainly constrained by the availability of feedstock. Thus we see significant increases in production for scenarios where feedstock availability is high. Grain-based ethanol is hampered by higher feedstock prices and competition with food markets, which leads to declining volumes in the long term. Cellulosic biofuels hold great promise if the necessary technology advances are made and these fuels can be produced at competitive prices. The potential for global cellulosic biomass production is sufficient to ensure that the resource base is not a constraining factor in the medium term, although the ability to bring the biomass to markets might limit access to these resources.

The results from the ETP global model show biofuel production dominated by the U.S., Brazil, and Europe. Combined, they supply more than 90% of all biofuels in 2010, although this decreases over time to about 70% in 2030. The majority of biofuel production is in the form of ethanol throughout the period. Sugar-based ethanol gradually loses market share as resource limitation prevents growth at the average rate of the industry.

Biofuel demand is highest in the same regions, but the South American countries are surplus producers and supply most of the internationally traded biofuels. In the reference case, the region exports 8 B gal of biofuels in 2020 and 12 B gal in 2030. The bulk of these volumes are sold to the U.S. and Europe, who are the largest importers.

The ESIA RFS is an ambitious policy that mandates 36 B gal of biofuel consumption in the U.S. by 2022. The challenge to the industry is vast and the scenarios we analyzed indicate that it may be difficult to reach the goal according to the schedule set out in the RFS. Developing the cellulosic resource base, building cellulosic biofuel production facilities, and constructing the ethanol distribution infrastructure quickly enough are the main obstacles to meeting the RFS. In the scenarios modeled, the shortfalls range from 0 to 4 B gal in 2020.

A hypothetical scenario that allows blending of up to 20% ethanol into gasoline[5] illustrates the infrastructure barriers that face ethanol when only 10% blending is allowed. This scenario also shows the competition between cellulosic ethanol and BTL where the former has a cost advantage when ethanol infrastructure constraints are not included and the latter has a cost advantage when ethanol infrastructure is a constraint.[6]

The main constraint to cellulosic biofuel production is infrastructure development rather than the underlying economics; thus, additional incentives such as growers' payments or an extension of the ethanol blenders' tax credit do little to increase overall biofuel supply. The blenders' tax credit would be paid to marketers for volumes already mandated by law and as policy tool would be inefficient to encourage biofuel supply. If increasing biofuel supply is the main policy goal, a targeted subsidy for cellulosic biofuels would have a larger impact and also be less expensive to implement.

In markets without biofuel mandates, the price of biofuels is determined by the price premium it can achieve over gasoline or diesel due to subsidy regimes. Higher oil prices will therefore lead to a stronger price signal for biofuel production and consequently the high oil price scenarios show higher worldwide demand for biofuels. In markets with mandates however, demand volumes are fixed through policy and changes in price signals do little to raise or lower demand. This is the case in the U.S. where biofuel demand is not very responsive to changes in oil price, because the buy-out from the cellulosic biofuels mandate adjusts to oil price and there are no relief-valve mechanisms for the other mandated volumes. This means that higher oil prices tend to lead to domestic production substituting for imports, because the oil price increase raises biofuel demand and thereby results in stronger competition in international markets.

A carbon price has a similar effect to that of a higher oil price. The carbon price can in fact be seen as a price premium on fossil fuels. A carbon policy will thus promote the production and use of biofuels worldwide. While higher oil prices are neutral as far as feedstock and conversion technologies are concerned, a carbon price will favor cellulosic and sugar-based biofuels production over grain-based production, which has higher carbon emissions per gal. A carbon policy will thus tend to increase the share of cellulosic biofuel and sugar ethanol at the expense of grain-derived ethanol.

In comparisons with the EIA's 2008 AEO biofuel projections under the new RFS, this study shows larger biofuel imports and therefore, larger U.S. supply. This is explained by two differences. First, this study focused on an in-depth analysis of the feedstock potential of biofuels exporting countries. Therefore, these results show a much larger supply of imports compared to the AEO. Second, the AEO assumes that the cellulosic technology is adopted at a much slower pace due to higher capital costs, whereas this study includes in its assumptions the provisions of the 2008 Farm Bill through a cellulosic production tax credit that reduces the cost of cellulosic biofuels production until it is competitive with corn ethanol production. This study also shows the benefits of U.S. policies to develop and commercialize cellulosic biofuels technologies since technology transfer results in the spread of cellulosic technology throughout the world.

This study provides an insight into why the U.S. is projected to have difficulty in meeting all of the mandates of the new RFS. The reference scenario, as well as analysis in EIA's *2008 Annual Energy Outlook* (AEO), shows that the U.S. is not projected to completely satisfy the biofuels requirements of the new RFS in 2015. A compliance gap is estimated to persist through 2030 in EIA's analysis. Our reference scenario shows a significantly smaller compliance gap through 2025. There are three main constraints to meeting the requirements of the new RFS: 1) ethanol infrastructure cost, 2) limits to cellulosic biofuels production growth in the early years of technology development, and 3) large demand operating at the inelastic limit of the sugarcane supply curve. These challenges were discovered through the use of various policy and market scenarios.

INTRODUCTION

Record high oil prices as well as concern over potential threats to energy security and the global environment have made biofuels an urgent national priority (IPCC, 2007)[7]. In response, the President proposed and Congress enacted a greatly expanded Renewable Fuel Standard (RFS) in 2007. Title II of the Energy Independence and Security Act of 2007 (EISA) mandates a total of 36 billion (B) gallons (gal) of renewable fuels by 2022—more than five times U.S. 2007 production[8]. The RFS does not allow more than 15 B gal per year of starch-derived ethanol (e.g., corn ethanol) to satisfy the mandates, a figure that is not substantially higher than current and planned U.S. production capacity. Consequently, the RFS essentially mandates second- generation biofuels or biofuels with lower greenhouse gas (GHG) emissions than corn ethanol (and even lower emissions compared to the petroleum fuels they displace). Also, second- generation biofuels are not derived from food crops and consequently pose much smaller ecological risks (land use intrusion and biodiversity).

The Energy Information Administration (EIA) has projected that the mandate of 36 B gal of biofuels will not be met by 2022 (AEO, 2008). This global study extends EIA's analysis by estimating worldwide potential to produce and supply biofuels to the U.S. market in the EISA timeframe, and determining the constraints under which the RFS might not be met. While U.S. production of biofuels is more desirable from the perspective of the national security, biofuel imports (because they come from different parts of the world than oil) also increase fuel source diversity and U.S. energy security. Both domestic and international production of second- generation biofuels can benefit the global environment.

World biofuels potential depends on the amount of biofuels feedstock that can be grown on available land, biofuels production capacity, and the economic competitiveness of biofuels compared with other options. We identified eight regions[9] that have the largest resource potential to produce biofuels for export to the U.S. Supply curves for selected feedstocks[10] and conversion costs for selected technologies were estimated in each relevant country. The results were used to update the global Energy Technology Perspectives (ETP) model. The ETP model was then used to project the volume of biofuels on the world market through 2030 including production, consumption, and trade within and among the fifteen world regions of the ETP model (biofuels production was also estimated for several specific countries within the "Central and South American" ETP region). A range of scenarios were considered to examine the impacts of policies and market conditions on biofuels production and exports.

A key result of this assessment is the importance of promoting the rapid commercialization of cellulosic biofuels if the EISA RFS requirements are to be met. One important factor for promoting the rapid commercialization of cellulosic biofuels is the $1.01 per gal production tax credit (net) for cellulosic biofuels provided under the 2008 Farm Bill[11]. We conclude that it would be much less likely that the RFS requirements would be met without this production tax credit. Cellulosic biofuels are being encouraged in the U.S. as the mainstay of future biofuels supply because they are more sustainable in several respects than ethanol derived from grain. First, feedstocks for this form of production do not have an alternative use such as food or feed. Increased production would therefore not affect food markets directly, although competition for land, labor, and capital could generate indirect impacts. Second, the resource base is potentially enormous; and third, lower water and fertilizer requirements and a significant reduction in carbon emissions could make cellulosic ethanol more environmentally friendly than starch based ethanol.

Scenarios

Scenarios were developed to understand the impact of different policies, technology prospects, and market uncertainties on global biofuel production, imports, and exports. All scenarios modeled include the RFS mandates, as well as existing biofuels policies in other countries. In all scenarios the RFS mandate after 2022 is modeled such that the renewable fuel requirements grow in proportion to the projected baseline petroleum demand growth. All scenarios include the aforementioned cellulosic biofuels tax credit and, implicitly, many other Farm Bill programs which are likely to be reauthorized in the 2012 and subsequent Farm Bills. Thus, in the reference scenario and all other scenarios, the cellulosic tax credit is represented as a continuing but declining production tax credit until cellulosic biofuels are cost competitive with corn ethanol. The incentive applies to both cellulosic biomass-to-liquids (BTL) and cellulosic ethanol.

Reference Scenario: Similar to the AEO 2008, the reference scenario is a projection of the current policy and market environment. As detailed later, it includes all biofuels relevant policies in the U.S. and worldwide.[12]

Full descriptions of all scenarios can be found in Appendix A. Descriptions of key scenario variables follow:

Policy Scenarios: The policy scenarios explore possible U.S. and global policies:

- *E20:* This scenario assumes that E20 is a certified[13] fuel for conventional gasoline vehicles allowing up to 20% ethanol blending in gasoline.

- *$50 per t CO_2:* There is an implicit global carbon value that reaches $50 per tonne (t) of CO_2 emitted in 2030 (Figure 1).

- *Credit and Tariff Extension:* The blenders' tax credit of $0.51 per gal of ethanol and the U.S. tariff of $0.54 per gal on ethanol imports are continued indefinitely.[14]

Market Scenarios: These scenarios explore market possibilities that will influence biofuels production and trade.

- *High and Low Feedstock Availability:* These scenarios are based on historical variations in feedstock production as described in the sections of the report on Methodology and Feedstock Assessment Results.

- *High and Low Oil Price:* These scenarios include potential high and low oil prices (Figure 2). The oil prices are OECD import basket prices and will generally be significantly lower than spot prices for reference crude oils like West Texas Intermediate or Brent. The reference, high, and low prices are endogenous values from the model itself. The higher oil price reflects the prices currently prevalent in the market and is discussed as a separate scenario (Alfstad, 2008).

Sensitivity and Uncertainty

The model results are relatively sensitive to some of the key assumptions. One of the most important is the cost at which ethanol can be produced from cellulosic feedstocks. At the conversion costs reported in the NREL study (Bain, 2007), cellulosic ethanol is highly competitive and penetrates the market very rapidly. If this low production cost cannot be achieved, because the assumed technology investment due to the U.S. cellulosic biofuels subsidy does not achieve rapid technology learning, then cellulosic biofuels production will be less than estimated.

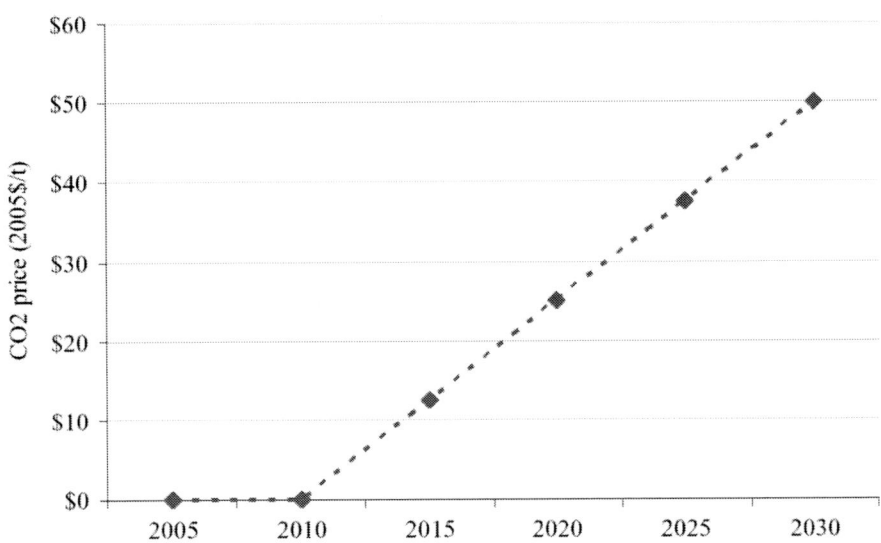

Figure 1. Global price per tonne of carbon dioxide in the CO_2 price scenario

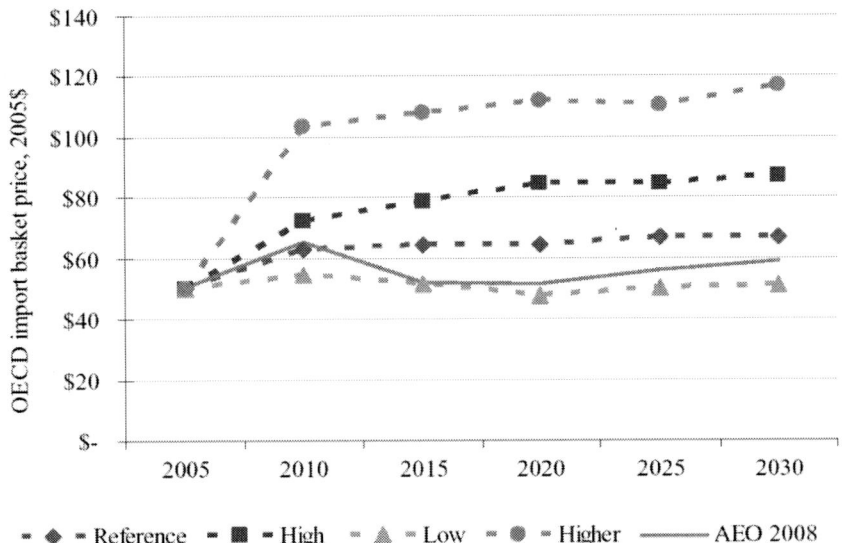

Figure 2. Price of oil in the AEO, reference, low, high, and higher oil price scenarios

There is also great uncertainty in the rate at which cellulosic ethanol can gain market share even if low production costs are achieved. The available historical data is insufficient to determine appropriate market penetration rates for this technology at different ethanol prices.

Assumptions regarding biofuels policies in Europe are also important, especially in early years. Europe attracts a large share of imports in early years because of high subsidies. The assumption here is that when the European Union (E.U.) goal under the current policy (10% by 2020) is met, the subsidies are no longer available and any additional ethanol is available on the global market. This assumes that Europeans are willing to pay these subsidies even if a large share of their supply comes from abroad. It also assumes that infrastructure to accommodate these imports is put in place as well as clear customs regulation. If these conditions are not met, considerable amounts of additional biofuels (1-2 B gal) could be available to U.S. markets.

Another note on the European subsidy regimes is that they are generally in the form of tax exemptions. If biofuels achieve substantial market shares (which they do in all scenarios) this leads to a considerable loss of revenue. To make up for this loss, the tax exemptions either need to be phased out as they are currently in the ETP model (although not necessarily at the same rate), or European governments will have to find an alternative source of revenue to pay for roads. There is great uncertainty regarding these subsidy provisions as they generally expire after a few years and there is little clarity as to what will replace them. A faster phase-out than what has been assumed here is a distinct possibility, which could significantly impact trade flows.

Another issue regarding the representation of Western Europe in the model is that it is modeled as a single region. While the region has a range of fuel taxes and exemption levels, it is represented as a single market with one average fuel tax and one tax exemption level in the ETP model. This means that the model exhibits some "knife-edge" behavior, in which U.S. importers are either competitive against all European importers or not. In reality they could be competitive against some of the European markets but not others.

Reducing or removing tariffs and other barriers to international trade of biofuels would yield several benefits. Most importantly, it will allow the most cost-effective producers to expand production and more easily market their products abroad. This should reduce the overall cost of supplying biofuels to markets. It should also limit the political pressure to maintain subsidies provided these also benefit imported biofuels. A well established international trade network would also lessen the impact of bad harvests. If supply falls short in a region one year, countries with mandates can shop for biofuels in international markets. Risks are thus shared among a larger number of participants.

Environmental and Social Issues

Biofuel feedstock production can have negative or positive environmental and social effects depending upon the local situation and factors. These include the crop type, the methods used to cultivate and harvest the crop, and what the alternative land use would be. Biofuel feedstock production could lead to deforestation, and raise issues of land tenure, water use, and pollution, representing important and politically delicate issues in many countries. The capacity for longterm land use planning and enforcement is important to avoid or minimize the detrimental impacts of unsustainable expansion. Using food crops for biofuels production can also disrupt food markets.

Most countries in this study have established biofuels targets or mandates[15], partly in response to high crude oil prices. These targets provide investors with increased security based on assurances of local market demand. Many nations (including Brazil, Argentina, Colombia, and several CBI nations) also encourage investment through reduced tariffs and tax-credits. China and India have discouraged the use of food crops and prime farm land for biofuel production. Use of food crops, however, allows producing countries to build domestic biofuel industries and gear up for a transition to other technologies and feedstocks when they become available.

Carbon emissions related to biofuel feedstock production is also a contentious issue. Emissions from land use changes and from diesel combustion in farm equipment, water pumping, and production of fertilizer all erode the carbon benefits of biofuels. Different methodologies for estimating life cycle emissions produce different results. Most appear to arrive at estimates that show modest to significant emission benefits (Armstrong et al., 2002; Wang et al., 2007; Macedo et al., 2008; Sheehan et al., 1998). The net carbon loss of land conversion needs to be considered and is one of the greatest sources of debate and uncertainty. Soils and plant biomass are the two largest biologically active stores of terrestrial carbon and hold about 2.7 times more carbon than the atmosphere. If land is cleared to allow for cultivation of food or energy crops, the carbon contained in the standing biomass and some of the carbon stored in the soil will be released to the atmosphere. A "carbon debt" is thus incurred when native ecosystems are converted to cropland. This carbon debt is the difference between the amount of carbon stored in standing biomass and soil before land clearing and that of the crop grown in its place. The actual carbon debt is thus highly dependent on the type of ecosystem that is being cleared and the crop that replaces it.

One study (Searchinger et al., 2008) produced estimates for this pay-back. The actual pay-back will vary between scenarios, but even under their most optimistic assumptions it was over 30 years for corn, and under their base assumptions it was well over 100 years. Another study on the issue (Fargione et al., 2008) points out that the pay-back is highly dependent on the type of land being converted and the type of crop grown in its place. Their results indicate that clearing tropical or peatland rainforest to grow palm oil or soybean incurs large carbon debts with paybacks of several hundred years, while using abandoned or marginal cropland to grow prairie grasses incurs little to no carbon debt.

A recent World Resources Institute analysis (Bradley et al., 2007) of potential climate impacts of biofuels noted:

> While tropical production – as with Brazilian sugarcane or Southeast Asian palm oil – is energy efficient, there are significant carbon impacts from the land use changes that biofuels production demands. In the case of palm oil, both deforestation and the drying of peatlands (which release vast quantities of carbon when they burn) must be taken into account, and can overwhelm any emissions reductions from reduced fossil fuel use. In the case of sugarcane, this effect is less direct, as the sugarcane itself is not generally grown on newly deforested land. However, expanding sugarcane production creates competition with other land uses and puts further pressure on land availability, which, in turn, almost certainly results in carbon release from cleared lands.

In Europe, a proposal for a directive on the promotion of the use of energy renewable sources has been presented. This proposal calls for a binding 10% minimum target for biofuels in transport to be achieved by each Member State. In a public consultation of interested parties, the proposal suggested three sustainability criteria which were generally supported: a) land with high carbon stocks should not be converted for biofuel production; b) land with high biodiversity should not be converted for biofuel production; c) biofuels should achieve a minimum level of GHG savings (carbon stock losses from land use change would not be included in the calculation).

The EISA RFS contains specific requirements for the lifecycle GHG emissions[16] of each renewable fuel type (a minimum GHG reduction of 20% for all qualifying renewable fuel[17] compared to the fossil fuel it replaces; a reduction of 50% for advanced biofuels; a reduction of 60% for cellulosic biofuels; and a reduction of 50% for biomass-based diesel). Furthermore, this calculation must include "significant indirect emissions such as significant emissions from land use changes," as well as emissions from "all stages of fuel and feedstock production and distribution, from feedstock generation or extraction through the distribution and delivery and use of the finished fuel to the ultimate consumer, where the mass values for all greenhouse gases are adjusted to account for their relative global warming potential," thereby addressing the contentious issue of indirect land use. Feedstocks may include crops from previously cleared, non-forested land, biomass from private forest lands, managed plantations, algae, or separated yard and food wastes (Appendix B). The Environmental Protection Agency (EPA) will be releasing a notice of public rulemaking in the near future to determine which fuels would meet the GHG reduction and land use restrictions specified in EISA. Furthermore, the 2008 Farm Bill established a $1.01 per gal tax credit for cellulosic biofuels, incentivizing cellulosic biofuels, which have a smaller impact on land than crops.

Emission factors for feedstocks are shown in Table 1.

Table 1. Change in lifecycle GHG emissions per mile traveled by replacing gasoline or diesel with biofuels

Feedstock	Change from gasoline or diesel	Source
Sugar	-81%	Macedo et al, 2008
Corn (current average)	-19%	Wang et al., 2007
Cellulose	-86%	Wang et al., 2007
Wheat	-47%	Armstrong et al., 2002
Soy bean	-78%	Sheehan et al., 1998

The concerns discussed are observed in specific instances, and there appears to be a growing consensus that if best practices for socially and environmentally sound development can be applied, then appropriate biofuel feedstock crops could offer farmers enhanced employment and incomes (Kline et al., 2007). Such practices also could help reduce the burden of foreign oil imports on developing nations and avoid the worst potential negative impacts. Indeed, recent growth in biofuel feedstock production has been accompanied by greatly increased attention to what are often long-standing social and environmental challenges.

METHODOLOGY

This section describes the selection of countries for the study, generation of feedstock supply curves, and development of conversion technology costs. It also discusses the incorporation of the data for the selected countries into the ETP model[18].

Selection of Countries

Geographic areas were selected according to future biofuels production capacity for export to the U.S. market. The criteria used were:

- Current feedstock availability
- Current biofuels production
- Export infrastructure
- Processing capacity
- Proximity to the U.S.
- Forecast production potential

Seven countries and one region were selected as areas to be updated in the ETP model (Table 2). Data from a variety of sources were analyzed to identify the major biofuel feedstock crops in each country as well as other considerations affecting the feedstock supply available for conversion to biofuels[19].

Generation of Feedstock Supply Curves

For each potential feedstock in each area, Oak Ridge National Laboratory (ORNL) produced high, low, and base supply curves for 2012, 2017, and 2027. ORNL's base supply curves are projections of historic data within each country updated for the study. Changes in crop varieties, farming practices, weather, prices, government policies, and other variables led to historical variation in the area planted, yields, and total production. In the high case, growth is constrained to the upper limit of the trends based on historic yields and land use. The percentage of land allocated for a single crop in each state is not allowed to exceed 30% of that state's total land area, while the maximum yield in 2027 is set at twice the corresponding maximum yield reported in the U.S. The proportions of feedstocks allocated to food, fiber, and fuel in the model are based on historic fractions of production used to meet domestic needs.

ORNL compiled average production cost and projected supply data for each state in a country (or each country in a region) into supply curves from least to highest cost (see Figure 4 in Feedstock Assessment Results section)[20]. ORNL's report also discussed environmental, social, and policy considerations affecting, and affected by, feedstock production in each priority country. (Kline et al., 2007).

Table 2. Countries and feedstocks selected for study. A parenthetical checkmark [(✓)] denotes countries and feedstocks that have only a single data point, rather than a stepped supply curve projection. Cellulosic feedstocks also generally have limited price points

	Sugar, starch, and oil crops					Cellulosic feedstocks		
	Sugarcane	Corn	Wheat	Palm	Soybeans	Bagasse	Agricultural residue	Other
Argentina	(✓)	(✓)	✓		✓	✓	✓	✓
Brazil	✓	✓			✓	✓	✓	✓
Canada		✓	✓				✓	✓
China	(✓)	(✓)	(✓)		(✓)	✓	✓	✓
Colombia	(✓)			(✓)		✓		✓
India	✓					✓		✓
Mexico	✓	✓				✓	✓	✓
Caribbean Basin (CBI)	✓			✓		✓		✓
Share of world production (excluding U.S.)	73%	55%	27%	3%	81%	-	-	-

Table 3. Biofuel feedstock conversion technologies updated

	Sugar, starch, and oil crops					Cellulosic feedstocks		
	Sugarcane	Corn (dry mill)	Wheat	Palm oil	Soybean	Bagasse	Agricultural residue	Other
Fuel and technology								
Ethanol by conven-tional fermentation	✓	✓	✓					
Ethanol by bio-chemical conversion						✓	✓	✓
Ethanol by thermochemical conversion[22]						✓	✓	✓
Biomass-based diesel by transesterification				✓	✓			
Renewable diesel by hydrotreatment					✓			
Residual fuel oil by pyrolysis						✓	✓	✓
Biomass-to-liquids (BTL) by Fischer-Tropsch catalysis						✓	✓	✓

Development of Conversion and Transportation Costs

Updated transportation and country-specific conversion cost estimates were prepared by the National Renewable Energy Laboratory (NREL) for the feedstocks and technologies shown in Table 3. For most inputs, NREL used a statistical data reduction technique that accounts for observed variability among plants and the cost of each process used to convert a given biomass feedstock to fuel[21]. NREL also drew upon its previous process design studies (Spath et al., 2005), which included cost estimates for most components of typical facilities. Extensive discussion of this analysis and the data is presented in the NREL report prepared as part of this project (Bain, 2007).

Integration of Feedstock and Conversion Cost Update into ETP Model

Brookhaven National Laboratory (BNL) used the results of the feedstock and conversion technology studies to update the fifteen-region, global ETP model. For the countries studied, ORNL's biomass feedstock supply curves replaced or supplemented the ETP model's representation of biomass. Similarly, NREL's more detailed cost and performance data for biomass handling and conversion technologies, as well as for transport and distribution of biofuels within and among regions, replaced older, less complete data in the ETP model. As detailed in the BNL report prepared as part of this project (Alfstad, 2008), this updated ETP model was then used to evaluate the impact of technological progress and alternative fuel pathways on international energy markets under various scenarios.

General Description of the MARKAL/Energy Technology Perspectives (ETP) Model

The ETP model[23] was updated to include the feedstock and technology data gathered for this study. ETP is a fifteen-region, global MARKAL (MARKet ALlocation) energy model. MARKAL-based models are partial equilibrium models that incorporate a representation of a physical energy system. Figure 3 shows how components of the energy system are represented and linked together in a network, where the technologies form the nodes and are linked by flows of energy carriers. It has a representation of the flow of energy carriers through the physical infrastructure from the resource base through the various energy conversion technologies to the end-user. An unlimited number of technology types can be represented; therefore, the MARKAL framework allows a full comparison of technology options from resource extraction to service demand.

The MARKAL model is solved as a cost minimization problem where future states of the energy system are determined by identifying the most cost-effective pattern of resource use and technology deployment over time, discounted over the planning horizon. The MARKAL objective is thus to minimize the total cost of the system. Each year, the total cost includes the following components:

- Annualized investments in technologies
- Fixed and variable annual operation and maintenance costs of technologies
- Costs of exogenous (external) energy and material imports and domestic resource production (e.g., mining)
- Revenue from exogenous (external) energy and material exports
- Fuel and material delivery costs

- Welfare loss resulting from reduced end-use demands
- Taxes and subsidies associated with energy sources, technologies, and emissions

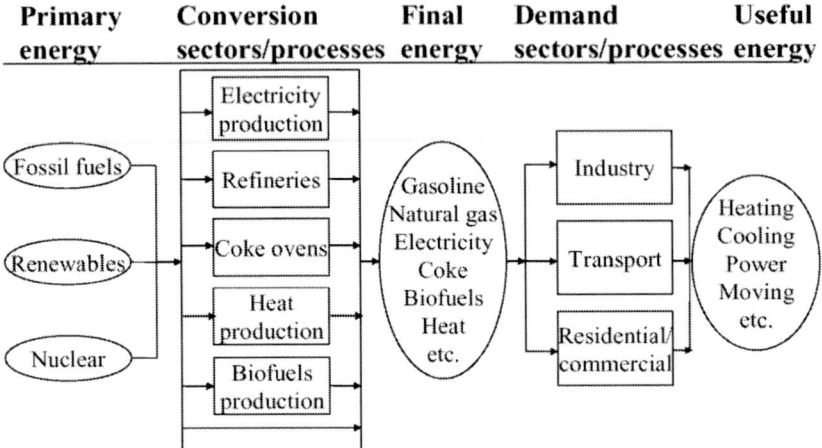

Figure 3. Energy flow diagram in ETP model

MARKAL-type models are demand driven, which means that for any feasible solution, exogenous demands (externally calculated and projected) are satisfied. The model then determines the least cost configuration of capital stock and utilization rates that will meet these demands, while obeying a set of user defined constraints such as natural resource availability, technology and capital availability, and environmental limitations.

The model is dynamic and tracks capital stock, so that the capital stock in any period is equal to the capital stock in the preceding period plus or minus any additions or retirements. The solution in one period is thus directly linked to the solution for other periods. Optimization is performed for all periods simultaneously, giving decision makers foresight or planning capability. As MARKAL models are defined by the performance, cost, and availability data for hundreds of energy technologies, any change in the data used for input parameters such as feedstocks or technologies will result in a MARKAL model that is different from its predecessor. Each region listed has a unique set of demands specified for all major energy services as well as energy intensive materials industries such as metals, ammonia, cement, and pulp and paper. Demands can be met with internal production or through trade with other regions.

The ETP database contains representations of hundreds of different technologies covering all stages of the energy system from extraction of primary energy to end use devices. This includes information on capital stock already in place as well as new technologies thought to be available now or at a future date. The general approach is that all technologies can be deployed in all world regions. To reflect real-world limitations, some of the following additional characterizations have been made:

- Region- and sector-specific constraints;
- Region- and sector-specific discount rates;
- Region-specific investment costs and fixed and variable costs;
- Region-specific supply curves for renewables;
- Region-specific lengths of seasons; and,
- Region-specific starting years.

As discussed above, the ETP model was expanded for this project. A new set of biomass supply curves developed by ORNL (Kline et al., 2007) for this study were added for selected feedstocks and countries (see Generation of Feedstock Supply Curves section above for details) in place of the existing representation. For Central and South America this also meant breaking up the aggregate supply curves for the entire region into sub-regional curves representing individual countries. The updated biofuels feedstock and technology data resulting from this study represents a significant update of the fuel sector in the ETP model.

The new representation is more detailed and yields smoother price response and behavior. Through interaction with the rest of the energy system these new supply curves can be used to predict the volume of feedstock that will be available for conversion to biofuels at a given price. Both volumes and prices can therefore be determined (endogenously) within the model. Cost curves for capital equipment have been included in the model. If the capital stock of a given technology is to expand faster than a predefined normal rate, a price premium must be paid for it. These new cost curves represent the added cost of outbidding competitors for labor, materials, and contractors. A given technology will thus take market share from competing technologies more rapidly if it has a greater cost advantage. These cost curves have been introduced for the various biofuel production technologies and are particularly important for the market penetration of cellulosic ethanol.

Policies Incorporated into the ETP Model Baseline

The principal U.S. policy covered in this study is Title II of EISA 2007 which mandates increasing amounts of renewable fuel use, reaching 36 B gal per year in 2022. This study focuses on the role of various biofuels in meeting this mandate. The U.S. also has a $1.00 per gal tax credit for producers of biodiesel. This study assumes that this tax incentive is extended throughout the study period[24]. In addition, with the passage of the 2008 Farm bill cellulosic biofuels receive a $1.01 per gal inclusive of any credits that may already be given, such as the credit of $0.51 per gal to blend ethanol with gasoline.

Since this is a study of global biofuel markets, policies that are in place in other countries were also accounted for. These policies are often in the form of tariffs and exemptions from fuel taxes. The policies modeled for key countries according to ETP defined regions are summarized in Table 4 below.

Table 4. Biofuels policies in key countries and regions in the ETP model[27]

Country/region	Gasoline tax (2006$/gal)	2010 biofuel tax exemption	Ethanol tariff	Other
Australia	1.40	100%	90 ¢/gal	
Canada	0.25	100%	20 ¢/gal	
China	0.15	100%	0	
Central and S America	0.70	50%	27 ¢/gal	Subsidy for hydrous ethanol and flexible fuel vehicles. Brazil blending requirement of 20-25%
Europe	2.80	90%	90 ¢/gal	5.5 % market share in 2010, 10% market share in 2020
India	1.90	0%	200%	500 million liters gasoline equivalent by 2010
Japan	1.85	90%	17%	
S Korea	3.00	90%	0	
USA	0.42	51 ¢/gal	54 ¢/gal	36 B gal of renewable fuels[25] by 2022. A declining $1.01 per gal[26] cellulosic biofuel credit is modeled until cellulosic biofuels become cost competitive with corn ethanol.

FEEDSTOCK ASSESSMENT RESULTS

The supply curves generated by ORNL for each country, feedstock, time period, and growth case replaced point estimates for supply and price in ETP. Table 2 (in the Methodology section of this report) lists the feedstocks studied.

As an example, Figure 4 shows the cost and production potential of sugarcane in Brazil in 2017 in the baseline growth case. Historic production trends and the structure of average production costs were analyzed by state (or province) to develop supply curves for each selected crop-country combination. Future supply was projected for 2012, 2017, and 2027 based on compound growth rates in yields and area harvested by state over the past seven years. The methodology assumes that recent growth trends for yield and harvested area at a state level will continue into the future within a set of defined parameters. Additional details of the methodology as well as data and supply curves for all feedstocks considered are provided in the ORNL report (Kline et al., 2007).

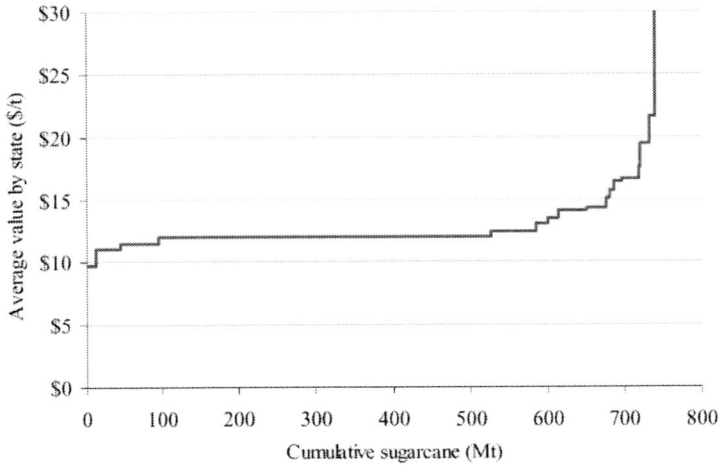

Figure 4. Supply curve example: Brazilian sugarcane in 2017 in the baseline case[28]

ORNL Study Findings

Supplies of all potential feedstock crops are expected to increase in the countries studied. Future growth in production is a result of increasing yields and expanding areas under cultivation. Feedstock growth rates vary widely, as does the portion of production available after meeting domestic demands for food, feed, and fiber. While one factor—the capacity to increase yields through improved crop varieties, technology, and production practices—is applicable to all feedstock crops and countries, the other factor, the capacity for expansion of area under cultivation, is limited to varying degrees across the countries studied.

Among the countries studied, Brazil has the greatest potential for expanded production, mostly from underutilized pasture. Much of its available, previously-cleared, and underutilized land is ideal for rain-fed sugarcane production. Argentina and Colombia also have relatively large amounts of underutilized arable land, along with capital and agricultural production technology which could enable them to quickly respond to policies and market signals for production. Figure 5 presents the 2017 projected available supply for export or biofuel use by feedstock in ethanol equivalent units, while Figure 6 presents similar data by country.

Sugarcane and soybeans provide more than 80% of the crop feedstock potential in the 2017 baseline case. The total crop feedstock potential is slightly smaller than the cellulosic feedstock potential in the 2017 baseline case[29]; however, there is greater uncertainty in the cellulosic feedstock potential compared to that of conventional feedstocks.

Conventional Feedstocks

Sugarcane production history and projections are presented in Figure 7. Other crops are presented in the report by Kline et al. Sugarcane output in the countries studied could grow from 999 million tonnes (Mt) in 2006 to about 1,460 Mt in 2017 and nearly 2,000 Mt by 2027 in the baseline[31] case. In the high growth case supplies could more than double by 2017, exceeding 2,000 Mt - ten years earlier than in the baseline case. Of current, widely cultivated crops, sugarcane has the highest yield of ethanol at the lowest cost and it represents over half of potential future supply available for export to global markets and/or conversion to ethanol over the next two decades. Growth in sugarcane production is led by Brazil where, compared to 2006 production, supply is projected to increase by 75% by 2017 in the baseline case and more than 130% in the high growth case. While Colombia and some CBI countries are expected to have similarly high growth rates in percentage terms, their total available supplies are small relative to those in Brazil.

The amount of sugarcane available for export or conversion to biofuels is only about half of the total supply because of its dual use as food sugar. Brazil is able to export more than three- quarters of its much larger production amounts while the CBI exports around half of its supply. Other countries export smaller shares of their production. In the baseline case, Brazil provides 86% of a total global projected available supply of 706 Mt, while CBI provides 6%, India 3%, and the other countries studied less than 2%. In the low growth scenarios, only Brazil and CBI are projected to have significant supply available for export or biofuel production, whereas in the high growth case, all sugarcane producing countries studied could contribute to biofuel supplies in the global market. Further, in the model it is assumed that Brazil, the country with an abundance of sugarcane feedstock, uses 60% of the sugarcane for fuel and 40% for sugar production. The only exception is in the high feedstock and increased Brazilian sugar ethanol scenarios where this ratio is 70 to 30. These cases (discussed in the report by Alfstad) assume that food sugar demand will grow more slowly than biofuel sugar demand, and that increasing sugarcane production will allow a larger share to be used for biofuels. Sugarcane production is restricted by climate and soils to certain geographic areas, which is why Argentina, Canada, and the U.S. are not major producers. Brazil, by contrast, has available land suitable for sugarcane estimated at several times greater than the present land area used for this crop.

Soybeans have the second highest production potential of the feedstocks studied in the selected countries, particularly in Argentina, Brazil, and China. The baseline case estimate for total soybean production in the selected countries shows a rapid increase from 2006 levels of about 108 Mt to more than 200 Mt in 2017 – an 87% increase. The other 2017 estimates range from 147 Mt in the low growth case to 288 Mt in the high growth case. Most of the increase is in Brazil and Argentina, the world's two top exporters of soybeans and soy products, respectively. Given relatively limited domestic demand for soy in these countries, most of the increased production in these countries will be available for export markets and/or biofuel production.

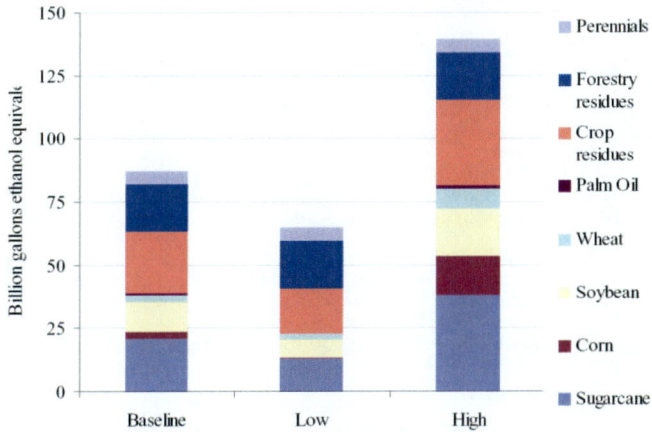

Figure 5. Feedstock available for export or use in biofuel production by source in 2017 in countries analyzed

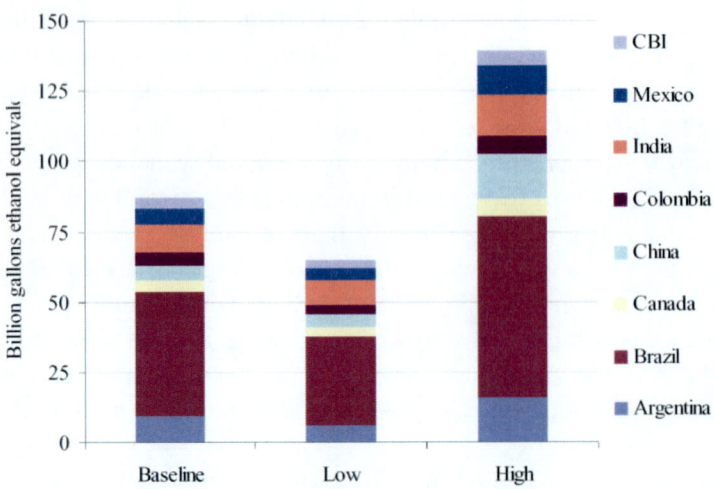

Figure 6. Feedstock available for export or use in biofuel production by country in 2017[30]

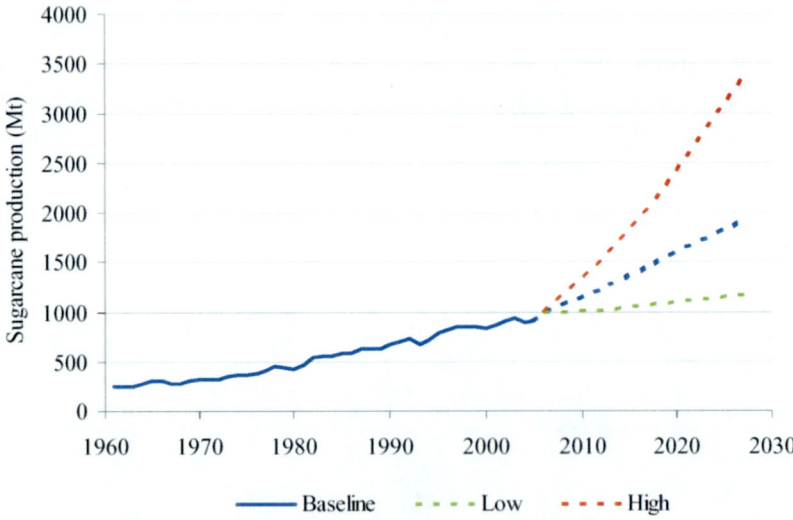

Figure 7. Actual and projected sugarcane production; total aggregate supply from countries studied

Corn production is projected to increase steadily in the five countries in this study that grow it (Argentina, Brazil, Canada, China, and Mexico). Corn output is projected to increase in the baseline case by 39% from 2006 to 2017, less than the increase for sugarcane or soybeans. Furthermore, after domestic demand for food and feed is accounted for, only a small amount of projected corn production is expected to be available for export or biofuel production; about 7% in the baseline case and 1-2% in the low growth case. Of the countries studied only Canada is likely to use corn as a primary ethanol feedstock.

Wheat output in the three countries studied (Argentina, China, and Canada) is projected to grow even more slowly than corn. The baseline case's growth of about 10% from 2006 to 2017 is similar to worldwide growth in recent years with declining or stagnant production in the nations studied due to poor weather, low relative prices, and government policies. The high growth case's growth rate is more in line with rates for other feedstocks, with potential for supply to increase by as much as 43% above 2006 levels. Thus the low to high range by 2017 is 136 Mt to 208 Mt. The historical supply from these three countries represents only about 27% of global, non-U.S. wheat production, but the factors that affect these countries' growth also affect other producers so this fraction stays relatively constant.

Wheat is primarily grown as a food staple and it has seen very limited use as a biofuel feedstock to date. An exception is in the western provinces of Canada where about 9% of the national supply of wheat could be used for ethanol by 2012 in the baseline case. Canadian wheat ethanol plants benefit from the availability of low cost, lower quality (downgraded) varieties of feed wheat that might otherwise go to waste. The majority (72% in 2006) of the wheat supply in the countries studied comes from China, where a policy against making biofuels from food is likely to prevent any future ethanol production from wheat there.

Palm oil is by far the most rapidly growing biofuel feedstock in this study, though the share of total supply from the countries studied (Colombia and CBI region) is small relative to other feedstock crops. By 2017, palm oil production in Colombia and CBI is projected to increase by 150% over 2006 levels in the baseline case, and by as much as 250% in the high growth case.

Cellulosic Feedstocks

Cellulosic feedstock can be derived from three broad categories of resources including:

(1) Recoverable residues from field or plant processing operations, where availability can be calculated from the projected feedstock crop supplies in this study
(2) Waste and biomass associated with current forestry and fuel wood supply activities
(3) Potential dedicated energy crops – perennials which can be harvested regularly once established. The output of such biomass crops is a function of the estimated productivity of the species and arable land availability.

Given the scarcity of data for the latter two categories, the projections for the countries studied are very preliminary. Estimated cellulosic feedstock availability within the range of costs assumed in the study (generally under $100 per dry t) sum to 488 Mt, with just over half of this total derived from crop processing residues, primarily bagasse. Sugarcane-ethanol and palm oil biodiesel are the two fastest growing biofuel sectors identified in the study and their on-site endowments of low or no cost biomass wastes could facilitate the transition to cellulosic-based production with relative ease in the future.

Bagasse – the crushed stalk residue from sugarcane processing – is by far the most important single cellulosic resource identified for the countries studied, representing about 75% of all available agricultural crop residues in 2017 and about 40% of total cellulosic supplies estimated in that year. More importantly for a cost minimizing analysis, bagasse accounts for 78% of the low-cost cellulosic supply (241 Mt) that costs less than U.S.$36 per dry t (2017 baseline case). This is because bagasse is readily available at sugar-ethanol refineries and therefore (along with similarly available but smaller supplies of palm oil processing wastes and on-site wood mill residues).

The fact that a waste product such as bagasse costs anything is due to the fact that its use as a biofuel feedstock is a lost opportunity for its use as a fuel (direct combustion for process heat and/or electricity) and as a fiber. Most bagasse is currently burned as boiler fuel for sugar processing, but the expected steady increase in efficiencies of boilers over the coming decade means that an increasing portion of bagasse may be allocated to other uses, including biofuel, at an even lower cost. The use of bagasse for ethanol will also contribute to improving sugar- ethanol plant profitability.

Palm plant residue – As with sugar ethanol plants, palm oil processing plants already generate substantial cellulosic waste. With palm, however, the cost as a feedstock is likely low or even negative as the residue has a volume of 1-2 times palm oil output. Palm plant wastes already far exceed thermal process needs and even present disposal costs, making them ideal candidates as a biofuel feedstock. However, palm wastes do not contribute significantly to the biofuel feedstock potential of the countries included in this study. This resource would be more important to consider in other countries with significant palm production, for instance in Southeast Asia.

Table 5. Biofuels conversion processes included in the analysis

Source	Feedstock	Conversion	Product	Distribution/Consumption
Sugarcane	Sugar	Sugar-ethanol		New distribution Infrastructure required Consumption limited to E10 for most of existing U.S. vehicle fleet Higher blends (i.e. E85) can be used in a currently small portion of the U.S. fleet
Corn Wheat	Starch	Dry mill	Ethanol	
Bagasse/other agricultural residue		Biochemical conversion		
Forestry residues	Cellulose	Thermochemical conversion		
Energy crops		Fischer-Tropsch synthesis	Distillates, naphtha	Products are refining feedstocks Compatible with conventional fuel
Oil palm Soybean	Oil	Transesterificiation	Biomass-based diesel (fatty acid methyl esters)	Can be blended with diesel at high ratios in most areas of the U.S.

Other Cellulosic Supplies – Estimated costs for other cellulosic supplies vary significantly depending on assumptions related to productivity and collection and transportation costs. These other feedstock supplies generally have estimated average prices above $36/dry t. These include sustainable recovery of

- corn stover;
- wheat straw;
- fuelwood;
- wastes associated with industrial forestry; and
- perennials[32] – dedicated energy crops harvested for biofuel.

CONVERSION PROCESSES AND FUEL TYPES

NREL developed a complete set of plant gate price curves for each conversion technology in each country in the study (Bain, 2007). Table 5 describes the range of technologies and fuels integrated into the ETP model using results from the NREL study, as well as Fischer-Tropsch synthesis costs already in the ETP model.

NREL Study Findings

For biofuels derived from crops, feedstock cost is generally the largest piece of the plant gate price and shows the most variation among countries, while capital and non-feed operating costs show little variation among countries (Figure 8). For the more complex and capital intensive cellulosic conversion technology, capital and operating costs are proportionately larger, but feedstock costs may still be an important factor in determining cost-effectiveness. A comparison of thermochemical[33] cellulosic ethanol plant gate prices is presented in Figure 9.

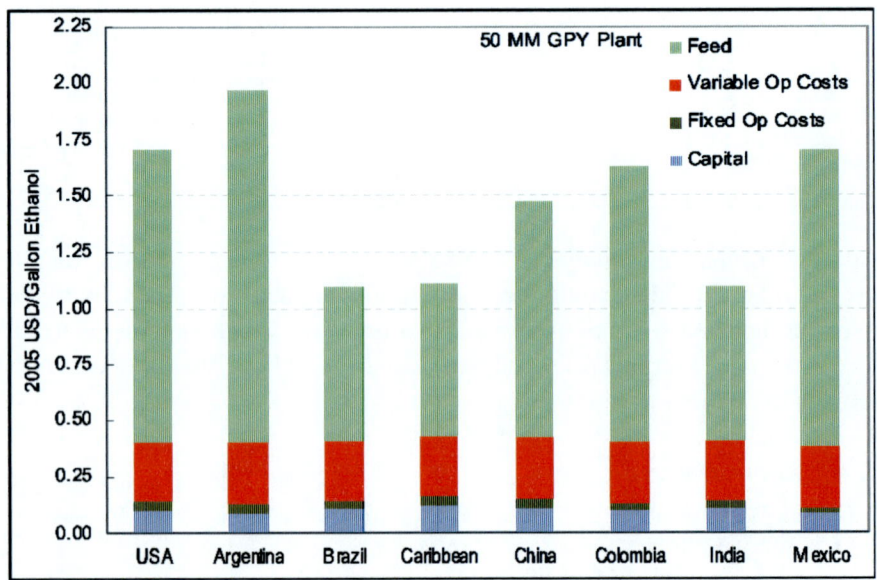

Figure 8. Comparison of sugar cane ethanol plant gate prices[34] (Bain, 2007)

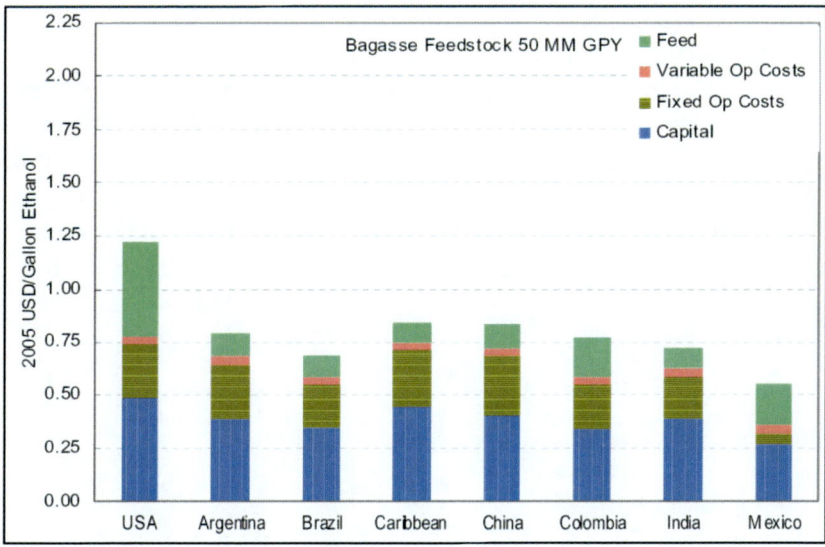

Figure 9. Comparison of thermochemical cellulosic ethanol plant gate prices (Bain, 2007)

RESULTS AND DISCUSSION OF SCENARIOS

The following results are intended to inform the biofuels policy debate by exploring the dynamics of the biofuels market and the relative impact of various policy and market uncertainties. This study is not intended as a forecast of future biofuels supply. The ETP model solves in five-year time periods; therefore, projections for 2022 are inferred from model results for 2020 and 2025. Detailed results are available in the BNL report (Alfstad, 2008).

World Biofuels Supply

Worldwide, we project 54 billion (B) gallons (gal)[35] of ethanol-equivalent biofuels production in 2020 and 83 B gal in 2030 as shown in Figure 10. The 2020 production projection is a four-fold increase over 2005. In all of the model years, biofuels from the U.S. and Brazil account for more than half of world production. For the range of production in each model year, the high value represents a scenario with high feedstock availability combined with a high oil price, and the lower value represents a scenario with low feedstock availability combined with a low oil price.

Consumption of biofuels worldwide is dominated by the U.S., surpassing half of worldwide consumption by 2015 and growing by almost 10% per year from 2005 to reach a ten-fold increase by 2030 (Figure 11). Western Europe is the second largest consumer of biofuels, followed by Central and South America as a region. Brazil is a net exporter of biofuels and Chinese and Indian consumption is relatively small because their biofuels targets are assumed to not be met.

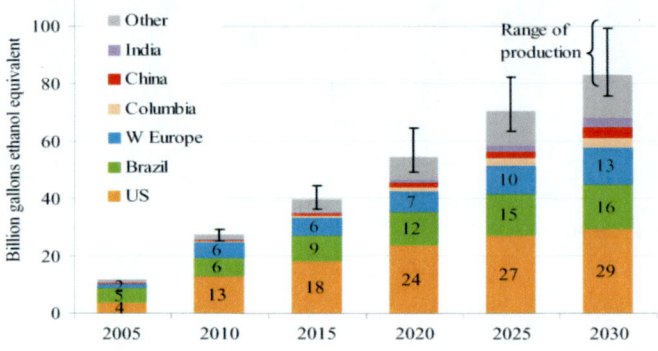

Figure 10. World biofuels production by country in the reference scenario

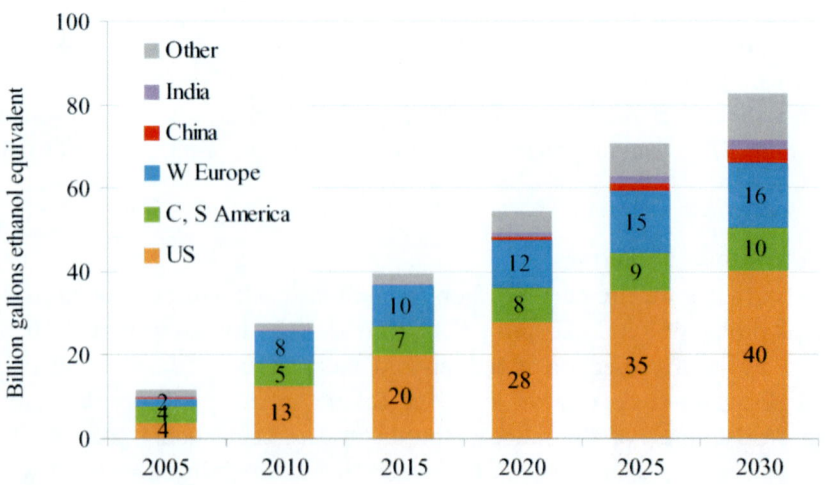

Figure 11. World biofuels consumption by country in the reference scenario

Grain and sugar-based ethanol dominate world biofuels production until 2020 when cellulosic biofuels begin to grow significantly in share (Figure 12). More than 75% of grain ethanol production occurs in the U.S., accounting for 15 B gal of the world production of 19 B gal (79%) in 2015; however, grain production levels off and declines by 2030 mainly due advanced renewable fuel requirements in the EISA RFS. Brazil is projected to maintain a greater than 80% market share of sugar ethanol production through 2030.

While domestic consumption is significant in Brazil, the country is capable of producing large volumes to satisfy world demand. Under the high feedstock availability scenario, Brazil can increase sugar ethanol production in 2020 from almost 10 B gal in the reference to over 16 B gal (Figure 13). While Brazilian sugar ethanol can satisfy the 4 B gal of the U.S. RFS advanced biofuels requirement that need not be biomass-based diesel fuel or cellulosic biofuels by 2022, if Brazil is to export larger volumes of fuel to the U.S. it will have to develop cellulosic biofuels production. Because of the ready supply of bagasse associated with sugar ethanol production, Brazil is well positioned to provide low-cost cellulosic biofuels, especially in light of the U.S. RFS requirement of 16 B gal per year of cellulosic biofuels by 2022. For this reason, Brazil and other countries will have a high incentive to invest in second-generation technologies to convert bagasse or other cellulosic feedstocks to biofuels.

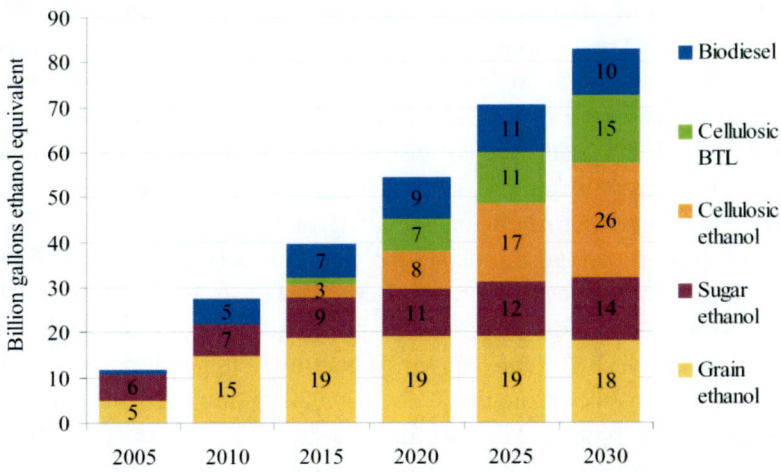

Figure 12. Total biofuels production by type in the reference scenario

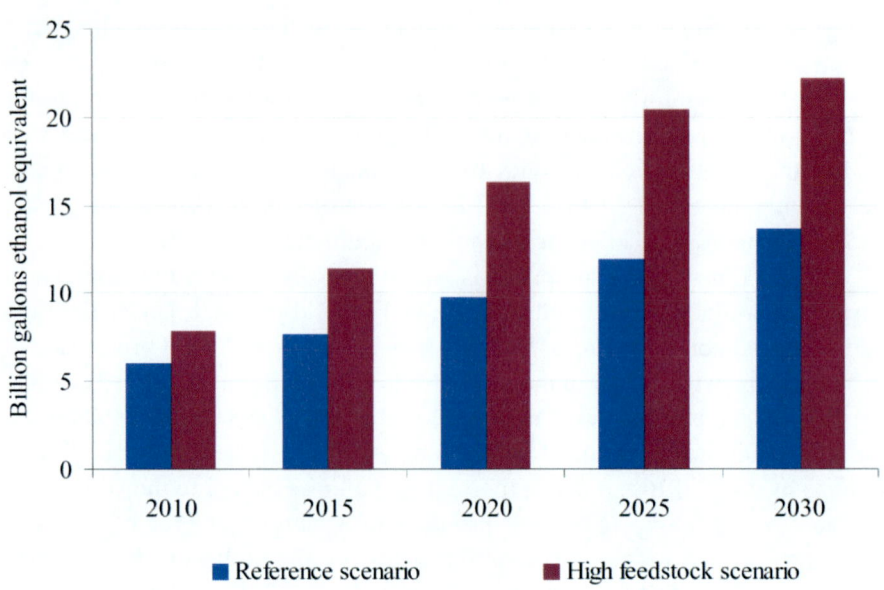

Figure 13. Brazilian ethanol production in the reference and high feedstock scenarios

Cellulosic biofuels are estimated to play an increasingly crucial role in meeting world demand for renewable transportation fuels. In the model, cellulosic feedstocks can produce ethanol via bio-chemical conversion as well as BTL via Fisher Tropsch catalysis. They are driven in particular by cellulosic advanced renewable fuel mandates in the U.S. RFS of 10.5 B gal by 2020, reaching 16 B gal in 2022. As a result of the demand set by this mandate, cellulosic ethanol production steadily increases in volume from the time the technology is adopted in the 2010-2015 timeframe and exhibits an average annual growth rate of almost 16% between 2015 and 2030 (Figure 12). With the demand created for cellulosic biofuels under the U.S. RFS, a significant portion of total biofuels supply in the latter years (2020-2030) is met through cellulosic technology. In 2030 more than 40 B gal of cellulosic biofuels (49% of total biofuel supply) are produced through either biochemical or Fischer-Tropsch BTL conversion. While more than 30% of cellulosic biofuel production occurs in the U.S. from 2015 onwards, there is also a substantial amount produced in countries that have an abundance of cellulosic feedstocks (such as Brazil). Investments in the U.S. (such as the tax credit in the 2008 Farm bill) to bring this technology to the marketplace leads to technology learning that can be transferred throughout the world.

Biodiesel is also an important component of world biofuel supply and increases to 10 B gal by 2030. A major portion of this is used in European countries where subsidies are the highest (Table 4).

U.S. Biofuels Supply

The main driver impacting biofuels supply in the U.S. is the EISA RFS requirement. The EISA RFS has specific mandates for each type of biofuel: renewable fuels, advanced renewable fuels, cellulosic advanced renewable fuels, and biodiesel advanced renewable fuels (described in Appendix B). Corn and wheat ethanol are assumed to qualify as renewable fuels and sugar ethanol is assumed to qualify as an advanced renewable fuel. Both cellulosic ethanol and cellulosic BTL are assumed to qualify as cellulosic advanced renewable fuels.

In the reference case, total U.S. biofuel supply undergoes significant growth to reach 28 B gal in 2020 and 35 B gal in 2025; however, this is less than the RFS requirement of 30 B gal in 2020 and 36 B gal in 2022 (Figure 14) with shortfalls in each of the biofuels categories except biodiesel and renewable fuel (e.g. corn ethanol). The EIA has also projected the impact of the new RFS on U.S. biofuels supply in its 2008 AEO. The AEO projects the RFS will not be met. It assumes that the U.S. Environmental Protection Agency exercises its authority to lower the mandate as necessary.

This study differs from the AEO in that it projects higher levels of biofuels supply in the U.S. due to more imports, reaching more than 35 B gal in 2025, whereas the AEO projects only 32.5 B gal by that year. The differences in these two models arise from two major underlying assumptions – the potential for other countries to export biofuels to the U.S. and the costs and availability of cellulosic biofuels. First, this study focused on an in-depth analysis of the feedstock potential of biofuels exporting countries. Therefore, our results show a much larger supply of sugar ethanol (mostly from Brazil) in U.S. biofuels supply (Figure 14) compared to the AEO. Second, the AEO assumes that the cellulosic technology is adopted at a much slower pace due to higher capital costs, whereas this study includes a cellulosic production tax credit similar to the Farm Bill provision[36] that reduces the cost of cellulosic biofuels production until it is competitive with corn ethanol production. This study also assumes that technology learning and transfer results in the availability of cellulosic technology throughout the world when economic. As a result, the cellulosic biofuels are produced in countries (such as Brazil) that have an abundance of cellulosic feedstocks and exported as cellulosic biofuels to the U.S. As expected, the AEO results show no imports of cellulosic biofuels, whereas this study shows larger volumes of cellulosic biofuels in U.S. supply when imports are added to domestic production.[37]

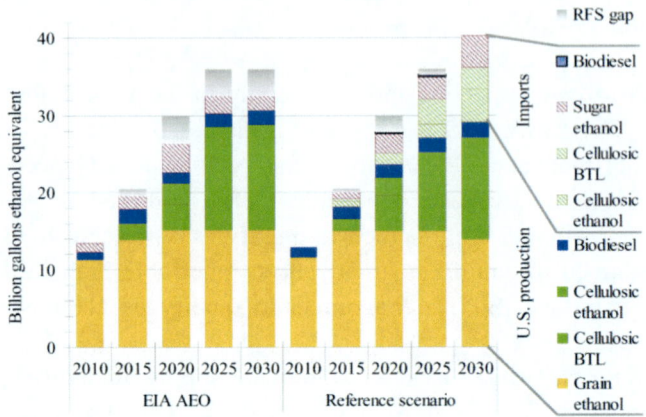

Figure 14. U.S. biofuels supply by type and source in the EIA AEO 2008 and the reference scenario

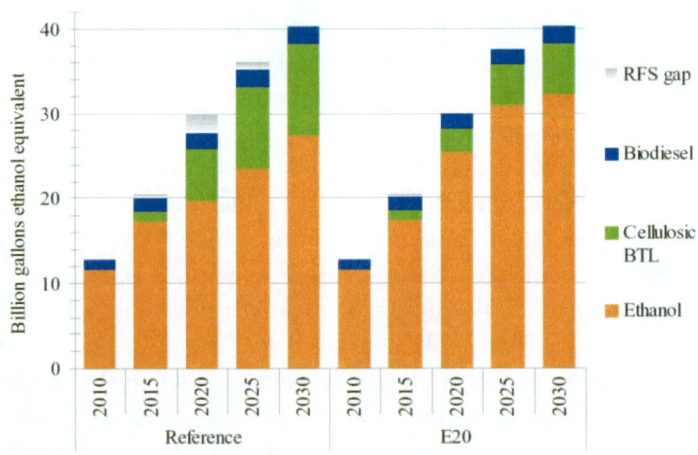

Figure 15. U.S. biofuels supply by type in the reference and E20 scenario

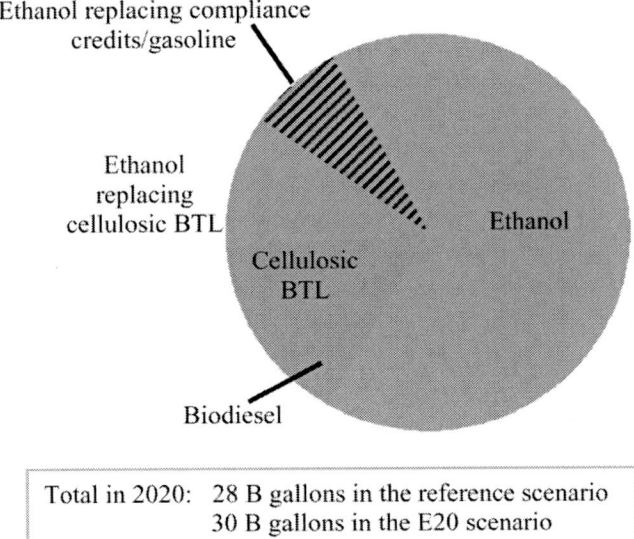

Figure 16. U.S. biofuels supply by type in 2020 in the E20 scenario

Various policy and market scenarios were modeled in order to understand why the U.S. may have difficulty in meeting the mandates of the new RFS.

Policy Scenarios

None of the policy scenarios in this study, with the exception of E20 scenario, significantly change the dynamics of biofuel production and imports to the U.S. As discussed below, the policy scenarios, with the exception of E20 scenario, have limited effect because the RFS is a strong policy in itself, ethanol faces infrastructure constraints, and feedstock demand (mainly for sugarcane) begins to operate at the inelastic regions of the supply curves.

E20 Scenario

The E20 certification scenario allows blending of up to 20% ethanol into gasoline. By assuming E20 certification, this scenario tests the hypothesis that ethanol infrastructure constraints prevent greater penetration of ethanol and attainment of the RFS. In this scenario ethanol infrastructure constraints are significantly reduced and more ethanol can be supplied to markets, allowing the U.S. to meet the mandates of the RFS. This not only leads to an increase in cellulosic and sugar ethanol but also a decrease in cellulosic BTL fuels. (Figure 15 and Figure 16)

Meeting the RFS not only requires the production of sufficient volumes of fuel with the available biomass but also the infrastructure necessary to bring this fuel to customers. Currently the U.S. blends up to 10% ethanol in gasoline. In 2020, gasoline demand is projected to reach 137 B gal (AEO, 2008) and the total ethanol required for blending in gasoline for E10 is much less that the mandated 30 B gal of biofuels under the RFS. To sell the mandated volumes retailers will need to market the excess ethanol as E85. Consuming large amounts of E85 requires dedicated infrastructure, including pipelines for transport of ethanol, fuel stations that can dispense E85, and flexible fuel vehicles to use the E85. Under an E20 scenario, these infrastructure requirements are not necessary because larger volumes of ethanol can be blended into gasoline using existing infrastructure.

Cellulosic feedstocks can be converted to synthetic distillates or naphtha via Fisher-Tropsch catalysis or ethanol via biochemical conversion. Both types of cellulosic biofuels are assumed to qualify as cellulosic advanced renewable fuels under the RFS. Cellulosic BTL fuels are more expensive than cellulosic ethanol (on an energy equivalent basis) but also more valuable because they can be refined and distributed with petroleum-based fuels in the existing pipeline infrastructure and can be used by the current fleet of light duty vehicles and trucks. Consequently, BTL fuels do not suffer from the same infrastructure constraints as E85. Suppliers (as represented in the model) will therefore prefer to meet the mandates with lower cost ethanol as long as it can be distributed at reasonably low costs. As the sales of ethanol increase, the cost of distribution goes up as the ability to expand ethanol infrastructure and sell E85 is limited. Above a certain volume of ethanol sales, the production cost advantage of cellulosic ethanol is negated by the escalation of distribution costs, and additional cellulosic biofuel supply will tend to be in the form of cellulosic BTL. These infrastructure constraints are the main reason for the significant supply of cellulosic BTL, as opposed to cellulosic ethanol only, in the reference scenario. Under the E20 scenario, infrastructure constraints on ethanol are relaxed, ethanol sustains its economic advantage over cellulosic BTL, and the mandated levels in the RFS are met. The E20 scenario is the only scenario where the RFS is met (Figure 15 and Figure 16).

$50 per t of CO_2 Scenario

This CO_2 policy scenario assumes a global implicit price of $50 per t of CO_2 (Figure 1). Under this scenario the requirements of the RFS are met after

2025 but not by 2020 (Figure 17), because this scenario has only a moderate effect on U.S. imports. This policy tends to provide a greater incentive for domestic consumption in producing countries as opposed to production for export. This policy also provides an incentive to increase the production of cellulosic biofuels and sugar-based ethanol as these emit less lifecycle CO_2 than biofuels from grain. As a result, U.S. grain ethanol consumption is lower than in the reference scenarios. The largest increase is for U.S. imports of sugar ethanol (almost doubling in 2025 compared to the reference scenario). U.S. cellulosic ethanol production increases by 40% between the two scenarios in 2025, but it is offset by a decline in imported cellulosic ethanol. Domestic and imported cellulosic BTL show small increases of 10% on average between scenarios in 2025. By 2030, cellulosic biofuels are a much more mature technology, and combined with its fewer CO_2 emissions than sugar ethanol, cellulosic biofuels growth outpaces that of sugar ethanol (Figure 18).

Credit and Tariff Extension Scenario

Extending both the blenders' ethanol credit and tariff has marginal benefits on overall biofuel supply. This policy protects domestic producers from foreign imports and U.S. production is boosted at the expense of foreign supply, with declines in imported sugar ethanol in 2020 and 2030[38] (Figure 19). Even on its own, the credit has limited impacts on biofuels supply and the tariff further diminishes the effect.

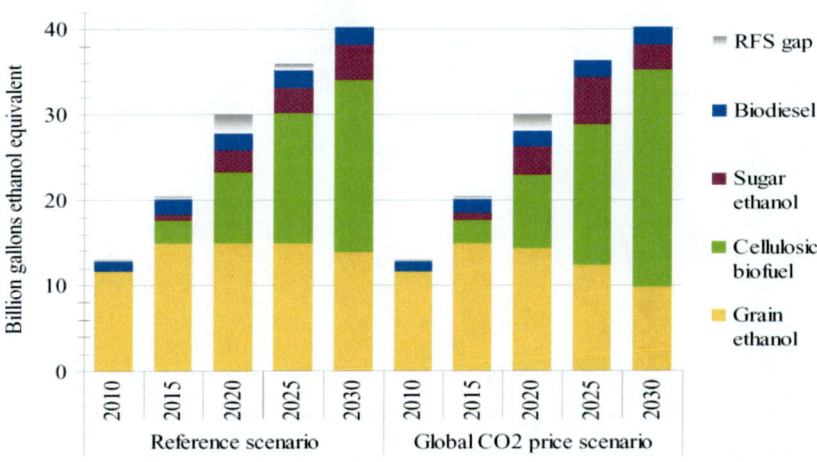

Figure 17. U.S. biofuels supply by type in the reference scenario and global CO_2 price scenario

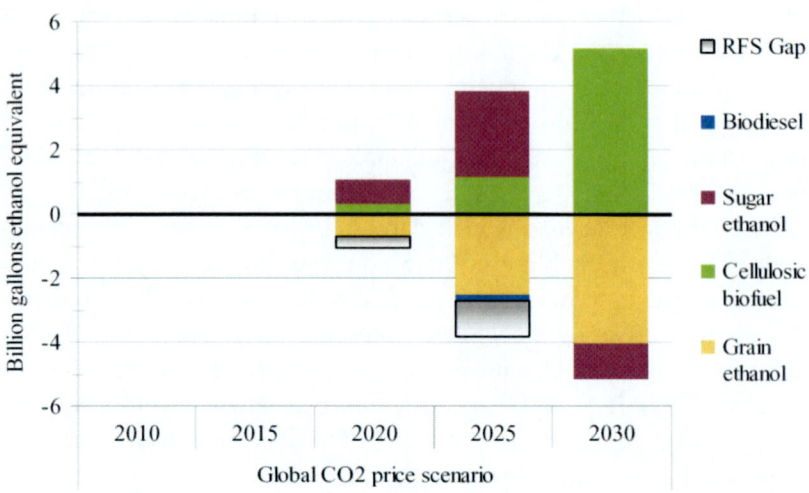

Figure 18. Change in U.S. biofuels supply by type in the global CO_2 price scenario compared to the reference scenario

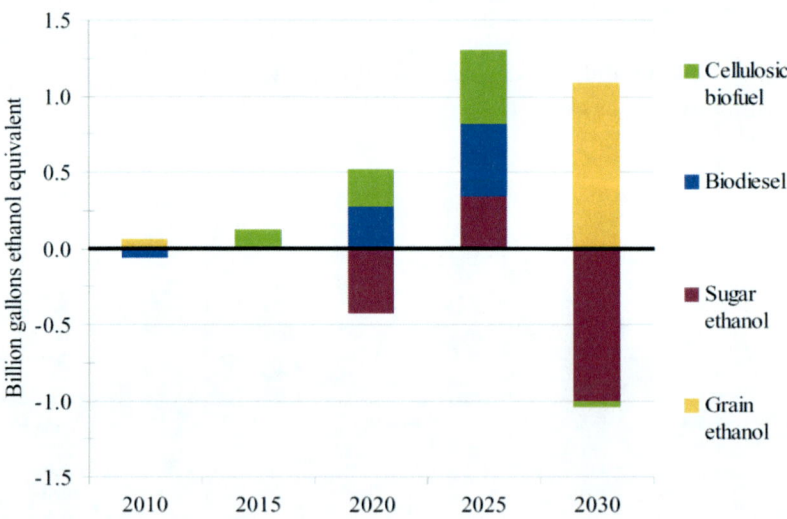

Figure 19. Change in U.S. biofuels supply by type in the blenders' credit and tariff scenario compared to the reference scenario

The blenders' ethanol credit is an incentive for a fuel that is limited by infrastructure constraints rather than poor economics. Corn ethanol production remains high and infrastructure constraints make it difficult to introduce more ethanol into the market, so the credit does not encourage significant amounts of additional cellulosic ethanol production. For corn and sugar ethanol, this scenario results in a shift in cost from consumers to taxpayers. For cellulosic ethanol, this scenario results in a transfer of funds from taxpayers to biofuel suppliers until the cellulosic advanced renewable fuel mandate in the RFS is met.

The RFS in itself is a strong policy so additional incentives do little to increase overall biofuels supply or cellulosic renewable fuels in particular. The main constraints for greater cellulosic biofuels production are the time needed to develop the cellulosic resource base, build production facilities, and improve the infrastructure for ethanol distribution, rather than the underlying economics. In the E20 scenario, where RFS mandates are met in 2020, E20 certification significantly reduces the ethanol distribution infrastructure constraints. The exact limitation on how quickly the infrastructure can be put in place is uncertain but is an issue that needs attention.

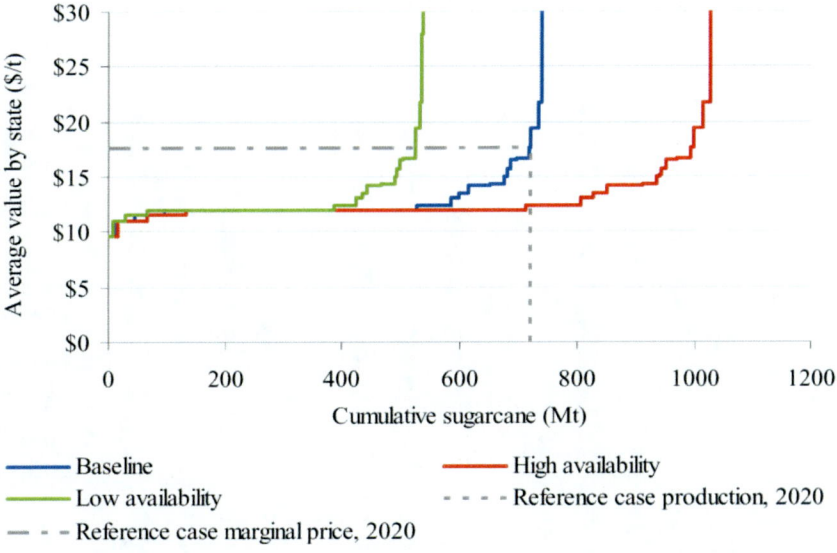

Figure 20. Brazil sugarcane feedstock supply curve in the baseline, low, and high feedstock availability cases in 2017. Superimposed is Brazilian sugar ethanol production in the reference scenario and the associated marginal price.

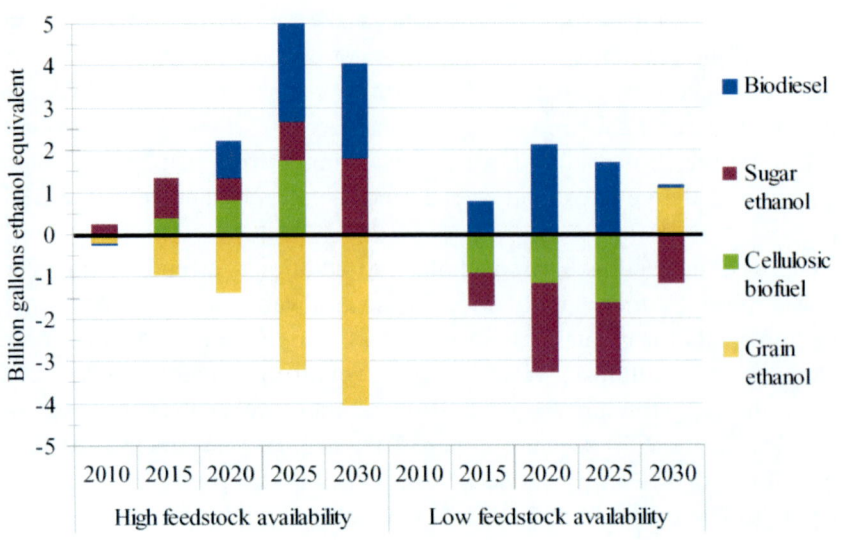

Figure 21. Change in U.S. biofuels supply by type in the high and low feedstock availability scenarios compared to the reference scenario

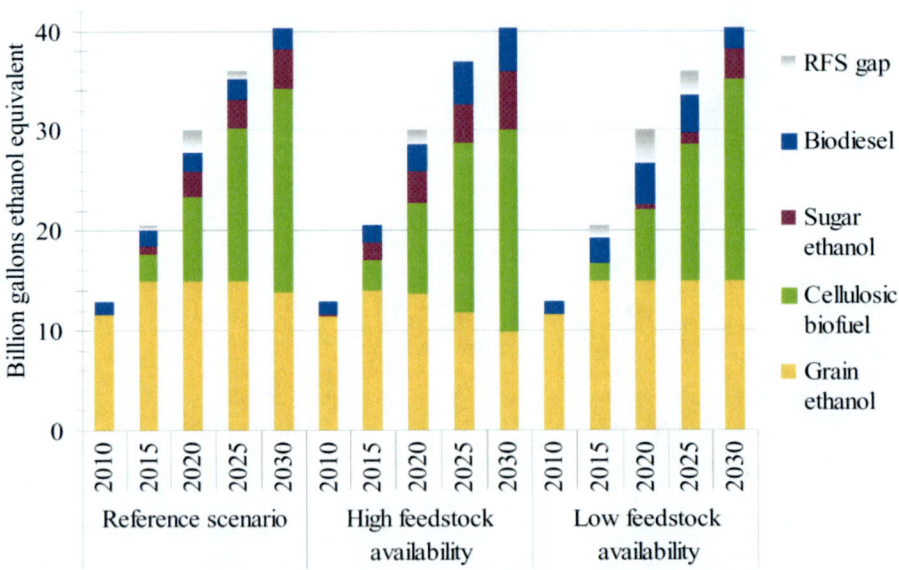

Figure 22. U.S. biofuels supply by type in the reference, high, and low feedstock availability scenarios

Market Scenarios

The market scenarios explore the market possibilities that could influence biofuels production and trade. In particular, this section explores the impact of high and low biofuels feedstock availability and high and low oil prices. Other scenarios are discussed in the BNL report (Alfstad, 2008).

High and Low Feedstock Availability

The high and low feedstock availability scenarios illustrate how sugar ethanol production in the reference scenario is mainly limited by its supply curve, where the large requirements of the RFS cause production to reach the inelastic portion of the curve preventing further growth in production at economic cost. Figure 20 illustrates how cumulative sugarcane production in Brazil in the reference scenario eventually reaches the inelastic portion of the supply curve where further increases in production result in significantly higher feedstock costs. This figure also shows how the high feedstock availability scenario relaxes this economic constraint and how the low feedstock availability scenario exacerbates it. In the high feedstock availability scenario, feedstock availability of sugar for both food and fuel is assumed to increase compared to the reference scenario and the ratio of fuel to food for Brazilian sugar application is assumed to be 70:30 compared to 60:40 in the reference scenario. As a result there are larger amounts of low-cost sugar ethanol available in world markets, along with cellulosic biofuels from the bagasse that accompanies sugar ethanol production, which displace significant amounts of domestically produced corn ethanol in the U.S. (Figure 21). The high feedstock availability scenario meets the requirements of the RFS by 2025 but not in earlier years. In the low feedstock availability scenario the opposite effect occurs; meeting the overall advanced biofuels target is much more difficult and costly (Figure 20).

High and Low Oil Prices

Changes in oil prices (Figure 2) have limited impact on U.S. biofuels supply because demand volumes are fixed through policy. Since the buy-out from the cellulosic biofuels mandate adjusts to oil price and there are no relief-valve mechanisms for the other mandated volumes, U.S. biofuel demand is largely unaffected by oil prices. The law removes the risk of falling oil prices to suppliers since the credit price will adjust to keep price incentives at levels that are presumed to be sufficient to ensure supply. Conversely, it reduces credit prices in the event of rising oil prices to ensure that the industry is not taxed needlessly in an environment where price incentives for biofuel supply should be adequate. Globally, high oil prices lead to higher demand for biofuels, which become more competitive with gasoline and diesel, particularly in subsidized biofuel markets. For the high and low oil price scenarios, U.S. renewable fuel consumption does not differ significantly from the reference scenario in quantity but does differ in the types of fuel consumed (Figure 24). Since higher oil prices increase biofuels demand in countries without fixed mandates, higher prices lead to a reduction in U.S. imports of sugar ethanol and a corresponding increase in U.S. production of grain ethanol (Figure 23). Low oil prices reduce demand in other world regions, increasing the supply competing in U.S. markets.

CONCLUSIONS AND RECOMMENDATIONS

This study projects that the U.S. may not meet the full mandates of the new RFS from 2015 through 2025, but the estimated gap is significantly smaller than that estimated by the EIA. There are three main constraints to meeting the requirements of the new RFS: 1) ethanol infrastructure costs, 2) limits to cellulosic biofuels production growth in the early years of technology development, and 3) large demand operating at the inelastic limit of the sugarcane supply curve. These challenges are illustrated through the use of various policy and market scenarios, and in particular the E20 certification scenario and the high feedstock availability scenario. The E20 scenario also shows the competition between cellulosic ethanol and cellulosic BTL where the former has a cost advantage when ethanol infrastructure constraints are not included and the latter has a cost advantage when ethanol infrastructure becomes a limiting factor. The results from scenarios like the global price on CO_2 also illustrate the interaction between grain ethanol and cellulosic biofuels which emit different amounts of CO_2. Results from the oil price cases reveal trade dynamics between producing and consuming countries and countries with and without biofuels mandates; however, both show little impact on U.S. biofuels supply because the RFS is a very strong policy in itself and mitigates the impacts of oil prices.

Further, in comparisons with the EIA's biofuels projections under the new RFS, this study shows larger biofuel imports and therefore, larger U.S. supply. This is explained by two differences. First, this study focused on an in-depth analysis of the feedstock potential of biofuels exporting countries. Therefore, these results show a much larger supply of imports compared to the AEO.

World Biofuels Production Potential: Understanding ...   79

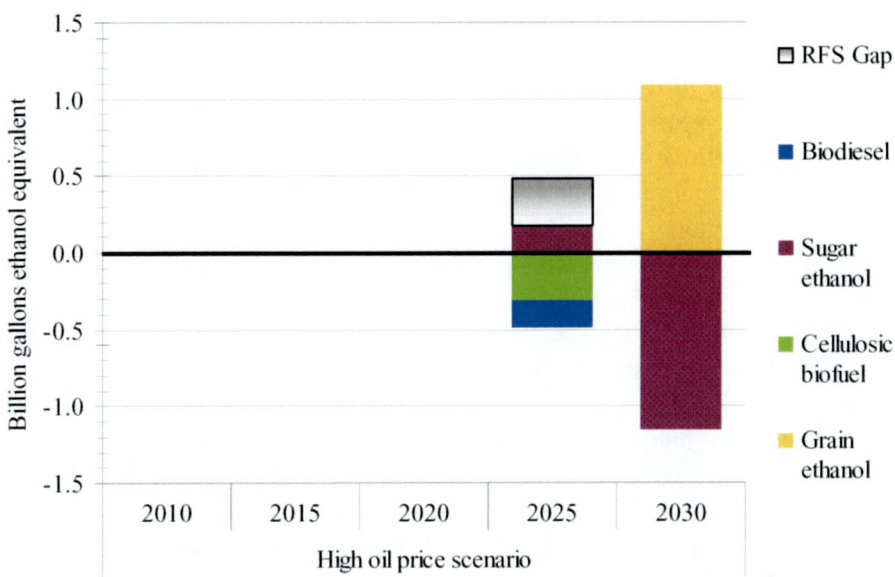

Figure 23. Change in U.S. biofuels supply by type in the high oil price scenario compared to the reference scenario

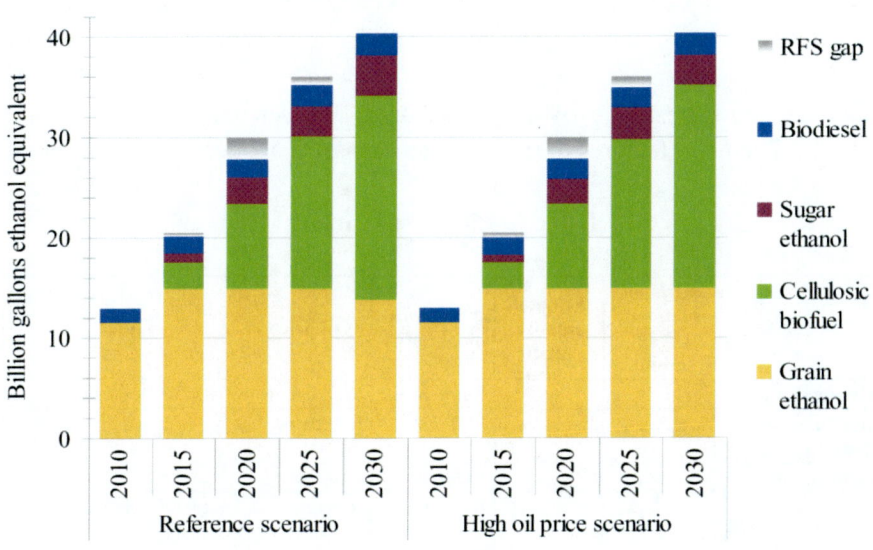

Figure 24. U.S. biofuels supply by type in the reference and high oil price scenarios

Second, the AEO assumes that the cellulosic technology is adopted at a much slower pace due to higher capital costs, whereas this study includes in its assumptions the provisions of the 2008 Farm Bill through a cellulosic production tax credit that reduces the cost of cellulosic biofuels production until it is competitive with corn ethanol production. This study also assumes that technology learning and transfer results in the availability of cellulosic technology throughout the world.

Further analysis is needed to assess biofuels production and export potential from countries not included in this study (regions of the world where original ETP biofuels data was used). This might not show any significant difference from our estimate of U.S. biofuel imports but would provide a better picture of how biofuels are produced and consumed throughout the world, especially in the Asian countries. Countries with biofuel production potential in Asia include Indonesia, Thailand, and Malaysia; and in Africa, countries such as South Africa, which have generally greater economic and political stability than other parts of the continent, could become major biofuel producers. New feedstocks that are not used for food, such as the jatropha plant, could be used to produce biofuels in these and other countries.

APPENDIX A: RESULTS FOR ALL POLICY AND MARKET SCENARIO

A detailed description of all the scenarios modeled is included in this appendix. The main features of the fourteen scenarios are described in Table 6.

Table 6. List of scenarios

Scenario	Ethanol blenders' tax credit	Ethanol import tariff	Feedstock availability	Oil price	Other
Reference case	-	-	Reference	Reference	60:40 fuel:food split
Credit and tariff extension	Extended	Extended	Reference	Reference	
Credit extension	Extended	-	Reference	Reference	
E20	-	-	Reference	Reference	Up to 20% blending of ethanol in gasoline is allowed
$20 per tonne (t) growers' payment	-	-	Reference	Reference	$20 per dry t of biomass feedstock

$50 per t of carbon dioxide	-	-	Reference	Reference	Global price of $50 per t of CO_2 by 2030
70:30 fuel:food split			Reference	Reference	70% of sugarcane in Brazil available for biofuels
High oil price	-	-	Reference	High	
Low oil price		-	Reference	Low	
Extra high oil price	-	-	Reference	Extra high	
High feedstock availability	-	-	High	Reference	70:30 fuel:food split
Low feedstock availability	-	-	Low	Reference	
High feedstock availability and high oil price	-	-	High	High	
Low feedstock availability and low oil price	-	-	Low	Low	

Scenario Analysis

Reference Scenario

The reference scenario forms the base against which other scenarios are compared. The reference scenario includes worldwide biofuel policies (see Table 4) including the U.S. new Renewable Fuel Standard (RFS). Particular attention was paid to the economic incentives required to commercialize cellulosic ethanol plants. We relied on the NREL cost estimates for these plants but accounted for the fact that these estimates were for the "nth of a kind" plant, not the first several plants. While the "nth of a kind" plant was estimated to be competitive, these earlier plants were not. Consequently it was necessary to assume some kind of learning investment that could come from a variety of public and private sources. The subsidy could be in the form of production tax credits, co-funding or other support or a subsidy scheme that improves the economics of cellulosic biofuel production for the investor. We had estimated, prior to the enactment of the 2008 Farm Bill, that an initial learning investment subsidy of about $1.00 per gallon[39] (gal) would be needed to achieve rapid uptake of cellulosic biofuels technology. In the later stages of our quantitative analysis, the 2008 Farm Bill was enacted which included a $1.01 per gal cellulosic ethanol production tax credit; this was included in our reference scenario. We assumed that the cellulosic ethanol production tax credit would be extended past 2012 at declining levels sufficient to keep new cellulosic ethanol plants competitive. We also assumed that the blenders' tax credit and ethanol import tariffs both expire in 2010.

Potential Future Policies

Since ethanol distribution infrastructure limitations and the ability to deliver sufficient volumes of ethanol to consumers is one of the main obstacles to reaching the RFS mandate, a policy scenario where E20 (gasoline blended with up to 20% ethanol by volume) is certified was considered. This would alleviate some of the infrastructure concerns, but the viability of this policy is dependent on resolving issues related to the impact higher blends of ethanol will have on engines and fuel systems. Currently most car manufacturers will only warrant their gasoline engines if they are fuelled with ethanol blends of 10% or less.

Another potential subsidy considered in this study is a growers' payment for U.S. farmers cultivating renewable cellulosic biomass. A $20 per dry tonne (t) payment would be offered to farmers starting in 2010 and expires in 2022. It is not adjusted for inflation.

Carbon Prices

In the carbon scenarios a carbon price of $50 per t of CO_2 is gradually phased in. It starts at $12.5 per t in 2015 and is increased by $2.5 per t annually until it reaches $50 per t in 2030. The carbon price applies to all sectors of the economy and to all regions. It is adjusted for inflation.

Oil Prices

Oil prices are determined endogenously in the model and are thus an outcome for a given model run and not an input assumption. Many factors influence the oil price, including supply curves, demand, end-use efficiency, and fuel switching. Another important variable is Organization of the Petroleum Exporting Countries (OPEC) rent-seeking, which can be manipulated in the model. This study used this rent-seeking behavior as the market driver for oil prices. A high oil price scenario is thus a case where OPEC follows a more aggressive policy and restricts supply to world markets by demanding higher economic rent on each barrel it produces. Low oil prices conversely occur when OPEC reduces their rent-seeking and produces more crude oil for world markets.

In this study high oil prices indicate an additional $18 per barrel economic rent sought by OPEC producers, while in the low price case it is $18 per barrel lower. It is worth noting that this does not mean that oil prices will be exactly

$18 per barrel higher and lower respectively for these two scenarios, since non-OPEC producers will respond to price changes. There is also an extra high oil price case where OPEC rent-seeking is raised by $43 per barrel. Oil price outcomes are shown in Figure 2.

The oil prices referred to in Figure 2 and in the rest of this report are the average U.S. imported prices of crude oil, also known as refiners' acquisition costs. Average import prices are generally significantly lower than the oil prices reported in the press. Reported oil prices are usually spot or future prices (i.e. WTI Cushing spot or NYMEX future price). These refer to reference crudes (light sweet crude oil), which are of higher quality than the average traded crude and thus receive a higher price. The difference between spot prices and contract prices varies widely, as shown in Figure 25. In its 2008 *Annual Energy Outlook* (AEO) the EIA projects this gap to range between $8 and $12 per barrel over the next 20 years (EIA, 2008a).

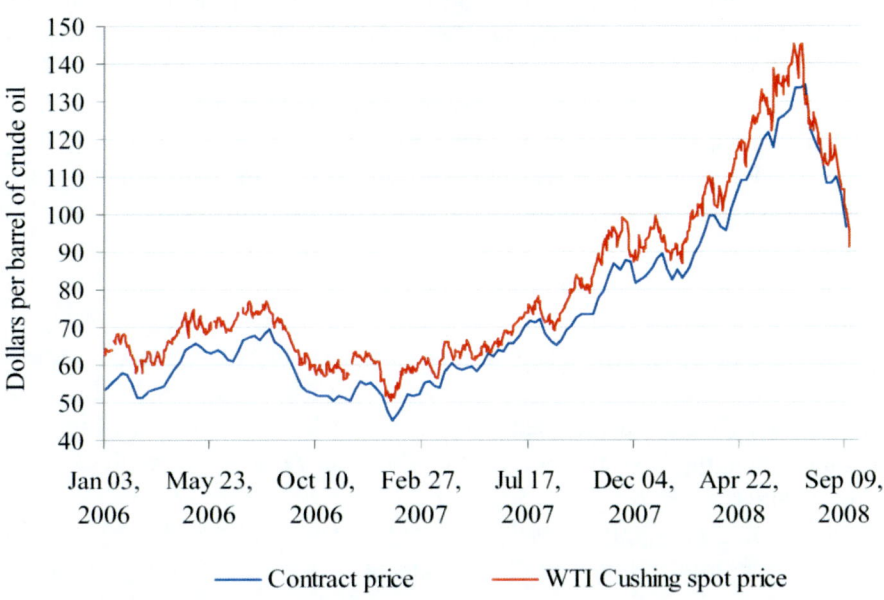

Figure 25. Historical oil prices by type

Results and Findings

Note on results and findings

Results from this study should not be read as forecasts. There are too many uncertainties and unknowns to make accurate predictions about future production and traded volumes of biofuels. This has been an exploratory scenario analysis that is meant to inform the biofuels policy debate. It is thus intended to address the dynamics of the biofuel markets and the relative impact of policies and market uncertainties rather than forecast future biofuel supply. Numbers should not be viewed in isolation but in the context of the study and underlying assumptions. For instance, an adjustment of supply curves would change produced volumes but the overall dynamics of the markets and the relative impact of the various policies should not change. In this section the focus is therefore mainly on these issues rather than on absolute numbers. That said, every effort has been made to ensure that the reported volumes are as reasonable as possible. Furthermore, the overall spread of outcomes from the analysis probably gives a fair estimate of the range of import volumes that are feasible in the medium term. This caveat is meant to emphasize the inherent uncertainty in this type of forward-looking exercise and encourage the reader not to attach too great an importance to individual values but rather view them in context of the overall range of results.

Results

This section covers the results of the scenario analysis. A total of fourteen scenarios were analyzed. To limit the page count and ensure readability, only selected data have been included here. Data tables for each of the scenarios can be found in the BNL report (Alfstad, 2008) for readers who wish to access the full results.

World biofuel supply for selected scenarios is shown in Figure 26. The reference case total biofuel production increases from 12 billion (B) gallons (gal) of ethanol equivalent in 2005 to 54 B gal 2020 and 83 B gal in 2030. The scenarios analyzed showed volumes ranging from 49 to 64 B gal in 2020 and from 75 to 99 B gal in 2030. The highest production worldwide occurs in the scenario with high feedstock availability combined with high oil prices. The lowest global production is found in the scenario with low feedstock availability and low oil prices.

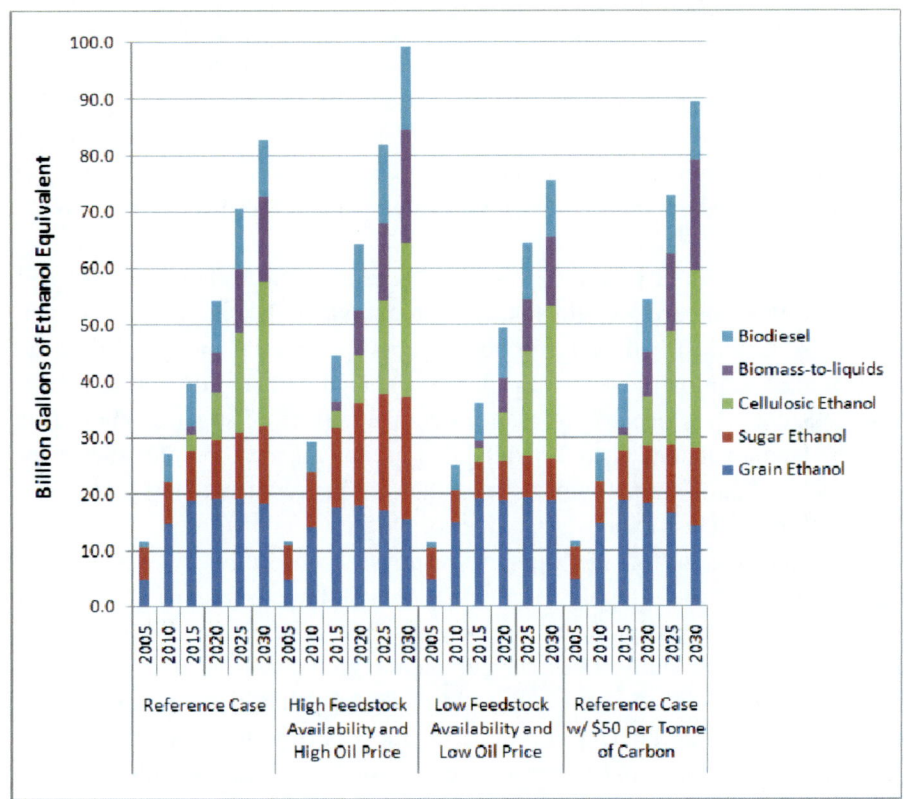

Figure 26. World biofuels supply by type for various scenarios

Initially, the majority of biofuels are produced from food crops. In the longer run, growth rates for grain and sugar ethanol slow down. This is mainly due to limits of feedstock availability, but also because the U.S. RFS does not mandate higher volumes for these fuels. Cellulosic biofuels quickly gain significant market share after they are introduced on a commercial scale in 2012. In the reference case cellulosic biofuels have a market share of 28% in 2020 and this grows to almost 50% by 2030.

Feedstock availability mainly impacts sugar ethanol and biodiesel production, as can be seen by comparing the high and low feedstock cases to the reference case. This is because grain ethanol is not competitive outside the U.S. and cellulosic expansion is not constrained by the overall resource availability under any scenario.[40] Again, it is worth noting that the scenarios exploring sensitivity to feedstock availability only adjust the supply curves for the countries covered in the feedstock assessment part of this study.

Worldwide sugar ethanol production in the reference case is about 11 B gal in 2020, while in the high feedstock growth case this rises to 18 B gal and in the low growth case it is as low as 7 B gal. The majority of sugar cane ethanol is produced in Brazil, which maintains a market share of more than 80% in all years for all scenarios. There is some feedback to grain ethanol production, as it is displaced when more inexpensive sugar ethanol is supplied to world markets.

Cellulosic biofuel production is more dependent on technology cost and limits to infrastructure roll-out than feedstock availability, as can be seen in Figure 26. Cellulosic ethanol production remains virtually unchanged by the shifts to high and low feedstock availability. It is in fact slightly higher for the low feedstock cases because of reduced competition from sugar ethanol. This is an indication that feedstock availability is not the constraining factor for cellulosic ethanol production but rather infrastructure constraints and competition from less expensive sources of biofuel.

Another reason for the lack of response from cellulosic biofuels to changes in feedstock supply is the fact that a large share of the overall cellulosic biofuel potential is in countries not covered by the feedstock analysis part of this study. Thus, the cumulative global shift in supply curves is smaller between scenarios for cellulosic feedstocks than for sugarcane or oil seeds, where a much larger share of total supply is from the studied countries (see Table 7). This is partly because cellulosic feedstocks are more evenly distributed geographically and because the country screening process mainly focused on potential for first generation biofuels (see section on Feedstock Assessment Results).

The introduction of carbon prices raises global biofuel production. Total biofuel supply is 7 B gal higher in 2030 after the introduction of a carbon price. The increase in cellulosic biofuel supply is higher at 10 B gal, while grain ethanol production drops about 4 B gal. There is also a small increase in sugar ethanol production. The reason for the low response in sugar ethanol production has to do with the shape of the supply curves. Sugar is the cheapest source of ethanol and supply is mainly constrained by feedstock availability.

Figure 27 shows the sugar cane supply curve for Brazil in 2030 in million t (Mt) produced at a given price per t. The carbon price results in a significant increase in the feedstock market price from $15.7 to $18.1 per t for the reference case. Since most of the ethanol is economic at prices below $15.7, we are at the inelastic part of the supply curve where response to price signals is relatively small. As a result the increase in production is quite limited from 1,020 to 1,067 Mt. The same argument explains why other price signals, such as higher oil prices, also have a limited impact on sugar ethanol production and why an outward shift in the supply curve (the high feedstock growth case) has a much larger impact.

Table 7. Share of world (non-U.S.) production of crop feedstocks represented by the assessed countries (Kline et al., 2007)

Feedstock	Countries Assessed in Present Study	2006 Output (Mt)	Share
Sugarcane	Argentina, Brazil, China, Colombia, India, Mexico, CBI	999	73%
Soybeans	Argentina, Brazil, China	108	81%
Corn	Argentina, Brazil, China, Canada Mexico	234	55%
Wheat	Argentina, Canada, China	146	27%
Palm Oil	Colombia, CBI	1.3	3%

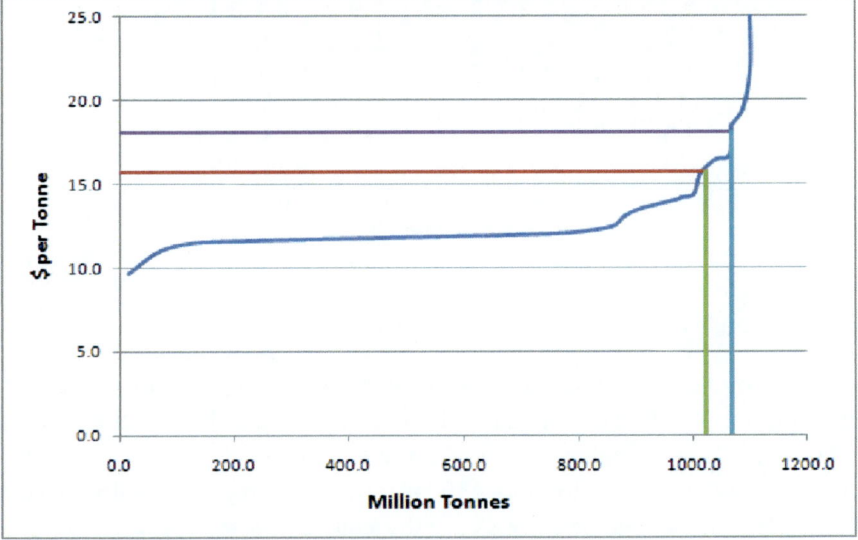

Figure 27. Effect of carbon price on Brazilian sugarcane supply in 2030

The reason why grain ethanol production is going down in spite of stronger price signals has to do with the limits on overall ethanol sales. U.S. and Europe, the biggest markets, are driven by mandates. The carbon price is not sufficient to encourage demand beyond the mandated levels and thus, competition is in a market of fixed size. Since the economics of cellulosic biofuels relative to grain ethanol improves, the former takes market share from the latter.

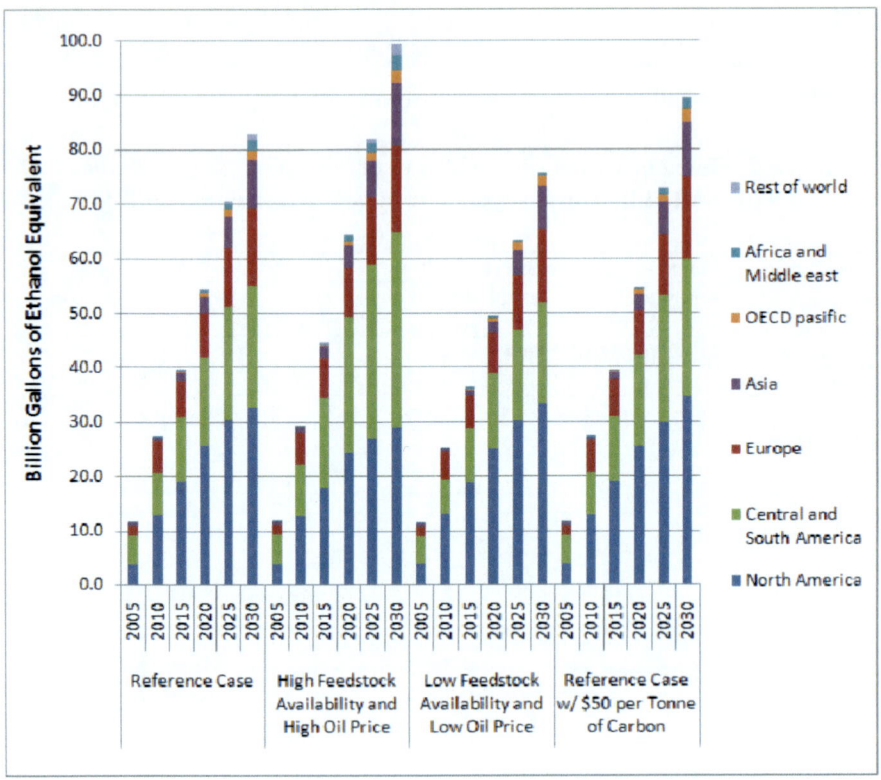

Figure 28. World biofuel supply by producing region for various scenarios

World biofuels production is dominated by North (primarily U.S.) and South (primarily Brazil) America, with significant contribution from Europe and Asia, as seen in Figure 28. In the reference case the U.S. share of total biofuel production grows to over 45% in 2015 but drops gradually to about 35% in 2030. In the high feedstock case the share drops to 26% in 2030.

Figure 29 shows biofuel demand by region. The U.S. is the biggest market for biofuels and it attracts around 50% of total supply for all years in the reference case. Europe and Brazil also attract large quantities of biofuels to satisfy their fuel standards. These three markets all have biofuel mandates, so overall demand changes little between scenarios, although there are some differences in U.S. demand since waivers can be purchased. The majority of the variation between scenarios is thus occurring in regions that have incentives but not mandates, such as Asia and Central and South America (other than Brazil).

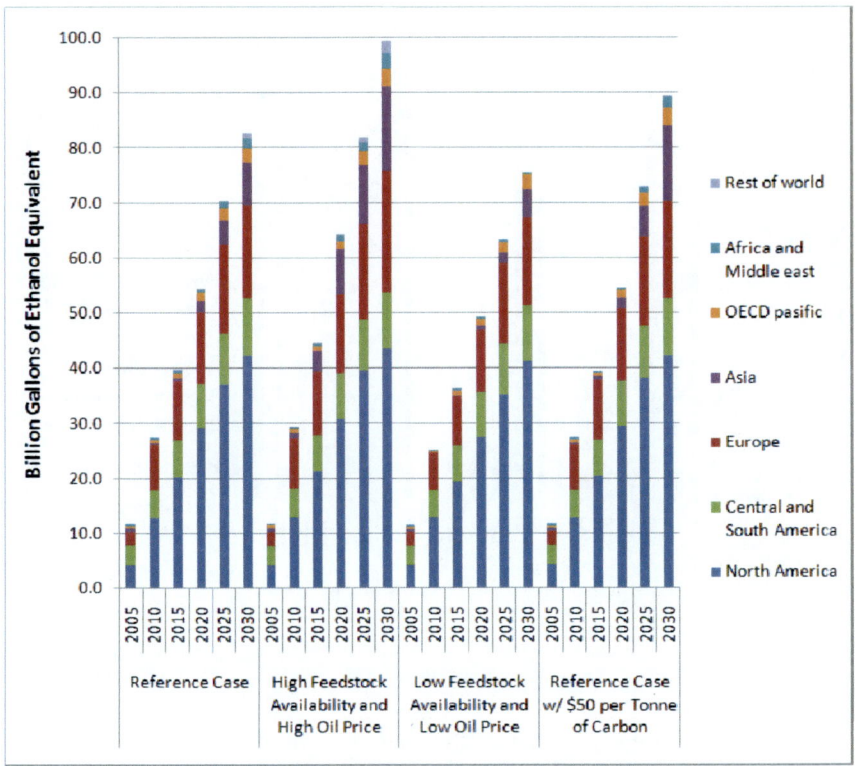

Figure 29. World biofuel demand by region for various scenarios

By comparing the supply side chart (Figure 28) with the demand side it is possible to develop a view of trade flows. A total of 10 B gal of biofuels is traded among regions in 2020 and this grows to 15 B gal by 2030. For the high feedstock growth case, these trade flows are significantly higher at 15 and 26 B gal respectively.

Most of the traded ethanol originates in Central and South America, which exports 8 B gal of biofuels in 2020 and 12 B gal in 2030. The U.S. and Europe are the major importers.

Figure 30 shows biofuel supply to the U.S. This supply is not sufficient to meet the cellulosic biofuel mandates in the early years of the RFS. As a result waivers are purchased to cover the shortfall. Total biofuel supply in the reference case is 28 B gal in 2020, of which about 4 B gal are imported. A little less than 20 B gal of this is ethanol, while the remainder is biomass-to-liquid (BTL) fuels (6 B gal) and biodiesel (1.9 B gal). Since the total RFS requirement for 2020 is 30 B gal, the waiver requirement is 2 B gal.

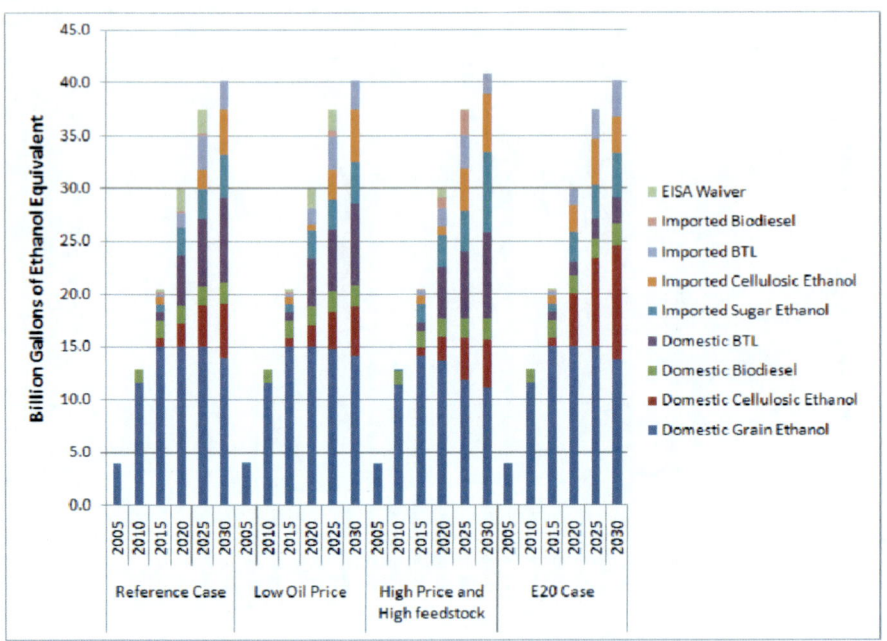

Figure 30. U.S. biofuel supply for the reference technology cases

Allowing E20 greatly increases the share of ethanol and displaces BTL. In the reference case ethanol distribution is restricted by infrastructure constraints and the inability to deliver sufficient volumes of ethanol as E85. This brings BTL into the market to fill the RFS gap. Distributing and selling ethanol as E20 alleviates these restrictions allowing more ethanol to penetrate the market. In fact, the E20 case is the only scenario where the mandate is met in 2020 and as a consequence some BTL is displaced. This is discussed further in the "biofuel market dynamics" section below.

In scenarios where there is abundant sugar ethanol available in the world market, significant volumes of U.S. domestic grain ethanol production are displaced. In these cases the advanced renewable fuels mandate is exceeded and the additional sugar ethanol demand is not driven by the RFS but by price. Extension of the ethanol blenders' tax credit or other domestic subsidy would protect domestic producers from this foreign competition but come at a significant cost to the U.S. Treasury.

The biodiesel requirement of the RFS is met in all scenarios in all time periods and supply also increases beyond the mandated volumes to help meet the overall advanced biofuels target. This happens in scenarios where sugar ethanol is in short supply and higher prices justify larger imports of biodiesel.

Biofuel Market Dynamics

The market adoption of biofuels worldwide is driven by a combination of mandates and economic incentives. If a mandate is in place the price will rise until supply is sufficient to meet it. Assuming that the targets set are achievable, there is no volumetric risk in this case, but there can be considerable risks associated with the cost of compliance. This cost risk can be mitigated through some form of relief-valve mechanism, but this would reintroduce volume risk. In a market where biofuels demand is driven by economic incentives (*e.g.* tax breaks, direct subsidies) the price of these fuels are determined by the price premium they can realize over gasoline or diesel due to the subsidy regimes. The cost risks are therefore reduced but there is significant volume risk. Therefore, under a mandate regime prices adjust to volumes, while under a subsidy regime volumes adjust to prices. In a region where both types of incentives are present the volumes will be determined by the mandates, while the main impact of subsidies will be to reduce prices for the consumer or increase profits for producers.

In a global market place, where some regions have mandates while others only have subsidies, these dynamics affect how biofuels are allocated among markets. Regions with mandates are willing to take the price risks and therefore tend to carve out a fixed share of the market, regardless of what happens elsewhere[41]. These markets would therefore tend to be served first. The regions that rely purely on policy induced price incentives will compete for whatever is available after all mandates have been met. Producers will sell their product in the market where they realize the highest net-back. The ethanol will thus tend to flow to the markets where price signals are the strongest, net of transport costs and tariffs. As long as there are unmet mandates this will limit the amount of ethanol going to countries without these fixed volumetric targets.

The EISA RFS creates a market place where ethanol is no longer a single commodity but can be separated into several subsets whose value is dependent on the feedstock from which it was produced. The ethanol itself will probably trade at one price but the associated credits will achieve different prices in the market place and thus change the total value of the ethanol. There are essentially four different types of ethanol under this regulation; ethanol that qualifies as renewable fuel (e.g. grain ethanol), ethanol that qualifies as advanced renewable fuel (e.g. sugar ethanol), ethanol that qualifies as cellulosic biofuels (e.g. cellulosic ethanol) and ethanol that does not qualify for any of the credits. It is thus perfectly possible, and in fact highly probable, that the market value (price plus credit value) for each of the types of ethanol will be different. Credit prices are generally determined by the market and will reach the level required to ensure sufficient supply, but for cellulosic ethanol (and other qualifying cellulosic biofuels) the credit value is equal to the waiver price as long as the mandate remains unmet.

The general dynamics of the biofuel markets as represented in this modeling exercise can best be understood by separating it into two stages; before the U.S. mandate is met, and after the mandate is met. Before the mandate is met all production is complementary in the sense that increased production in one part of the world does not come at the expense of production in another. There is little competition among ethanol producers, and ethanol prices are set by the price premium it can realize over gasoline due to the subsidy regimes. Any producer able to deliver ethanol to the market at this price will therefore do so. After the mandate in the U.S. has been met, biofuel markets become much more competitive and prices fall. At this stage (sometime between 2020 and 2030 in most scenarios), most of the subsidies in the rest of the world have been phased out. Producers are in more direct competition and increased production in one region leads to decreased production elsewhere. For instance, this is seen by comparing the high feedstock case to the reference case, where domestic production changes little in the early years but significant amounts are displaced by cheap imports in 2025 and 2030.

Another way to view this situation is to see the early years as a "sellers market" with high prices, where importers, unable to meet domestic biofuel targets, are competing for the fuel available on the open market. In the latter years the market is more balanced, prices fall and producers have to compete for market share.

Meeting the biofuel targets requires not only the production of sufficient volumes of fuel but also the infrastructure to bring this fuel to customers and consumers who actually purchase it. The biofuels mandate in 2022 is 36 B gal[42], roughly equivalent to 22 B gal of gasoline. In 2022 our projected gasoline demand is 139 B gal, so it is not feasible to meet the mandate by selling E10 alone. To sell more ethanol, retailers will be forced to market it as E85; however, sale of E85 requires a dedicated infrastructure.

Since the responsibility for meeting the mandate under the legislation lies with the "refineries, blenders, distributors, and importers," it falls on these entities to ensure that consumers actually purchase the fuel. This poses both infrastructure and marketing challenges. In order to sell the fuel to customers, there needs to be an underlying infrastructure to deliver it to them at fueling stations. Not only is an expansion of the distribution and fuelling station infrastructure needed to deliver sufficient volumes of E85 to customers, but there also must be enough flexible fuel vehicles on the road to use it. Approximately 1,200 out of almost to 170,000 fueling stations in the U.S. sell E85 currently (National Ethanol Vehicle Coalition, 2008). There are about 6.5 million flex-fuel vehicles on the road (EIA, 2007) but few of these actually run on E85.

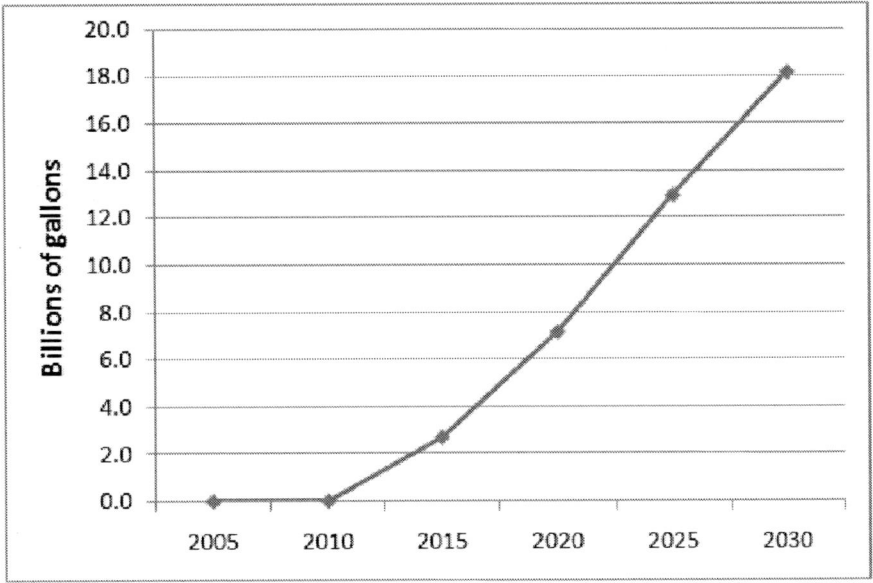

Figure 31. E85 sales in the reference case

Another concern is that consumers not only have to drive a flexible fuel vehicle and refuel at a station where E85 is available, they also need to be willing to buy the fuel. E85 currently trades at a premium to gasoline when adjusted for the loss in fuel efficiency due to the lower energy content for a given volume of fuel. Assuming that the bulk of consumers will buy the fuel that lets them travel at the lowest cost, this means that some form of additional incentive needs to be in place to encourage the adoption of E85. This is likely to require a cross subsidy.

Instead of supplying E85, marketers can meet the mandates by distributing cellulosic FischerTropsch BTL. These BTL fuels can be distributed with petroleum-based fuels in the existing infrastructure and can be used by the current fleet of light duty vehicles and trucks. Consequently, BTL fuels do not suffer from the same infrastructure constraints as E85. Based on the technology assumptions used for this study, cellulosic ethanol production is cheaper per unit of energy than Fischer-Tropsch diesel for a given feedstock price. The suppliers (as represented in the model) will therefore prefer to meet the mandates with ethanol supply as long as it can be distributed at reasonably low costs. As the sales of E85 increase, the cost of distribution increaases as the ability to expand the infrastructure and sell E85 is limited. Above a certain volume of ethanol sales, the production cost advantage is negated by the distribution cost escalation, and additional biofuel supply will then tend to be in the form of BTL.

This effect can be seen by comparing the reference case, and the credit extension case to the E20 case (Figure 32). There is little variation in overall ethanol sales in the two cases with E1 0 as the maximum blend. Additional incentives, in the form of a blenders' tax credit, have little impact. Since there are no substitutes for grain and sugar ethanol for the renewable and advanced renewable biofuel targets, the ethanol distribution constraints function as a cap on the amount of cellulosic ethanol that can be sold. Additional incentives for cellulosic biofuels will thus tend to lead to increased BTL supply. By allowing E20 to be sold to consumers, the infrastructure constraints are significantly reduced and more ethanol can be supplied to markets. This not only leads to an increase in cellulosic ethanol but also a decrease in BTL fuels.

The buy-out from the cellulosic biofuels mandate adjusts to oil price and there are no relief-valve mechanisms for the other mandated volumes; therefore, U.S. biofuel demand is not very responsive to changes in oil price. The RFS removes the risk of falling oil prices to suppliers since the credit price will adjust to keep price incentives at levels that are thought to be sufficient to ensure supply. Conversely, credit prices are reduced in the event of rising oil prices to ensure that the industry is not taxed needlessly in an environment where price incentives for biofuel supply are already adequate. Therefore, the price signal for U.S. renewable fuel suppliers does not differ significantly from the baseline to the high and low oil price scenarios.

While biofuel demand in the U.S. is relatively independent of oil prices, this is not the case in countries that use price incentives rather than mandates. In these regions biofuel prices are determined by the premium that the subsidy regime allows them to achieve over petroleum fuels. Lower oil prices therefore mean lower biofuel prices, which leads to reduced incentives to supply these markets. With lower prices and less biofuel going to international markets the cost of importing biofuels into the U.S. drops and volumes go up. Conversely, higher oil prices mean higher demand internationally, more competition for supply and therefore higher prices and lower volumes imported into the U.S.

This dynamic has two significant impacts. The first is the perhaps counterintuitive outcome that increasing oil prices lead to reduced biofuel demand in the U.S. This effect is marginal but might have been more pronounced if not for the constraints on ethanol sales described earlier. The second is that it changes the balance of imports and domestic production. Figure 33 shows that higher oil prices lead to reduced imports and thereby higher domestic production. The reverse would be true for falling oil prices. This relationship is consistent with the price impacts described above. This pattern changes if wholesale gasoline prices rise above $2.75, at which point the credit price decouples from the gasoline price and is held constant at $0.25. Oil prices above this level will increase biofuel demand in the U.S.

Critical to these outcomes is the assumption that producers believe that U.S. lawmakers will stick to this arrangement even if oil prices stay low for an extended period of time. The growth rates in biofuel production seen in these scenarios require strong confidence among investors that biofuel markets will expand and remain strong. Any doubt in the willingness of the U.S. and other governments to continue with their biofuel policies will likely lead to under-investment.

The RFS in itself is such a strong policy that additional incentives such as growers' payments or an extension of the ethanol blenders' tax credit does little to increase overall biofuels supply. The ability to develop the cellulosic resource base, build cellulosic biofuel production facilities, and construction of ethanol distribution infrastructure fast enough is the main constraint rather than the underlying economics. The exact limitation on how quickly the infrastructure can be rolled out is uncertain, and this is an issue that needs attention. The main impact of the credit extension is thus to bring in more domestic production at the expense of imports. Towards the end of the period, when growth rates are lower and infrastructure rollout is less of a concern, it also tends to bring in more cellulosic ethanol in place of BTL fuels. Furthermore, as more subsidies are introduced in the same scenario, diminishing returns to the cost of the policy will be seen.

The relative strength of the carbon value policy as a price signal compared to other policies changes over time. The blenders' tax credit and growers payment are both nominal values (i.e. they stay at a given dollar price and are not adjusted for inflation) and therefore decrease in real terms over time. The carbon price, however, increases over time and is also assumed to be in real dollars (i.e. inflation adjusted). This means that this policy strengthens over time.

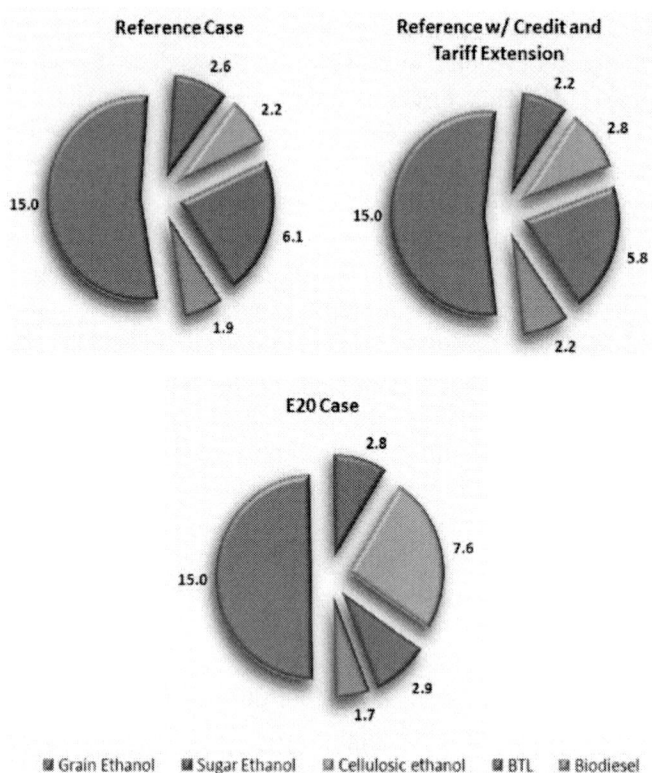

Figure 32. Biofuels sales by fuel type with reference scenario E10 versus E20

Countries in the NAFTA trade-zone such as Mexico and the Caribbean nations currently have preferential access to U.S. markets through import waivers. Producers in these countries can realize higher net-backs to the U.S than other exporting regions can for a given domestic ethanol price. The favored market for ethanol from these countries is therefore the U.S. and they tend to be the first exporters serving U.S. importers. An interesting consequence of this is that extending the current tariff and blenders' tax credit policies will lead to an increase in U.S. imports from these countries. This is because the blenders' tax credit will serve as an additional incentive to producers, while the tariff normally intended to cancel out the benefits of the tax credit to foreign producers does not apply to imports from NAFTA member countries.

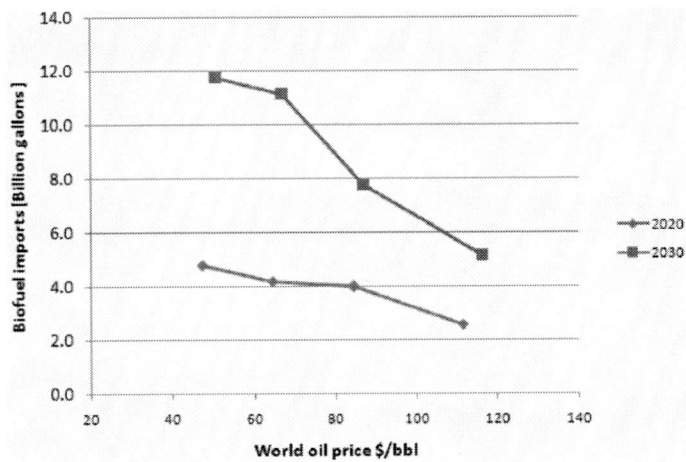

Figure 33. Oil price impact on biofuel imports

India and China are both potentially huge markets for biofuels but neither import much in most of the scenarios presented here. In the case of India this is mainly due to the very high import tariffs charged for imports of fuel ethanol into the country. Currently they are importing ethanol as a chemical feedstock but we assumed that this can be supplied by local producers when the internal infrastructure expands, which it will have to do if current ethanol blending targets are to be met. India thus produces significant amounts of ethanol for domestic consumption but does not import any. China does not currently have any strong economic incentives for ethanol imports. Fuel tax exemptions do not provide strong encouragement because the taxes themselves are so low. Like India, China relies mainly on domestic production in these scenarios. Any change in policy to help meet their ambitious targets might alter this.

APPENDIX B: ENERGY INDEPENDENCE AND SECURITY ACT OF 2007

The central policies covered in this study are the provisions enacted under the Energy Independence and Security Act of 2007 (EISA). EISA is designed to improve energy efficiency and increase the supply of renewable energy. The main provisions enacted into law can be summarized as follows:

- **Renewable Fuel Standard (RFS).** EISA mandates the use of additional renewable fuels by modifying the existing fuel standard. The standard now requires the sale of 36 billion (B) gallons (gal) of renewable fuels by 2022.

- **Corporate Average Fuel Economy (CAFE).** The law sets a fuel efficiency target of 35 miles per gal for the combined light duty vehicle fleet by the 2020 model year.

- **Appliance Energy Efficiency Standards.** The bill sets energy efficiency standards for a range of commercial and household appliances, including refrigerators, freezers, and lighting.
- **Repeal of Oil and Gas Tax Incentives.** EISA repeals two tax subsidies. The revenues from these taxes are intended to cover the cost of implementing the CAFE standards.

The focus in this study is on the RFS its impact on biofuel supply and demand. While the CAFE provisions have been included in the analysis, the issues examined in this study relate to biofuels and all scenarios are designed to address this subject.

Under the RFS provision the existing standard under the Clean Air Act is revised and expanded. The EISA amends the RFS in the Energy Policy Act of 2005 to include all transportation fuels. It expands the existing requirement to 9.0 B gal in 2008, increasing to 36 B gal in 2022. It requires renewable fuels produced at new facilities to have at least 20% lower lifecycle greenhouse gas (GHG) emissions than petroleum fuels. Starting in 2009, the RFS requires that an increasing amount of the above mandate be met using advanced renewable fuels, which are defined as biofuels derived from feedstocks other than corn starch with 50% lower lifecycle GHG emissions than the fossil fuel that it replaces. By 2022, it requires 21 B gal per year of advanced biofuel. Of the advanced biofuel mandate, specific carve-outs are made for cellulosic fuels and biomass-derived diesel[43]. (Figure 34) From 2015 onwards all increases in the overall mandated volumes are from advanced renewable fuels. The maximum volume of conventional biofuel (e.g. corn ethanol) for which distributors can receive credits is thus 15 B gal annually. The law does not actually prevent production of corn ethanol above this volume, but the economic incentives to produce it are significantly reduced if no additional credits are available. In general, compliance is required from "refineries, blenders, distributors, and importers" for each of the mandates volumes. The

Environmental Protection Agency (EPA) can issue waivers under certain circumstances and may also sell waivers for the cellulosic biofuel mandate at a price which is the higher of $0.25 or $3.00 less the wholesale price of gasoline.

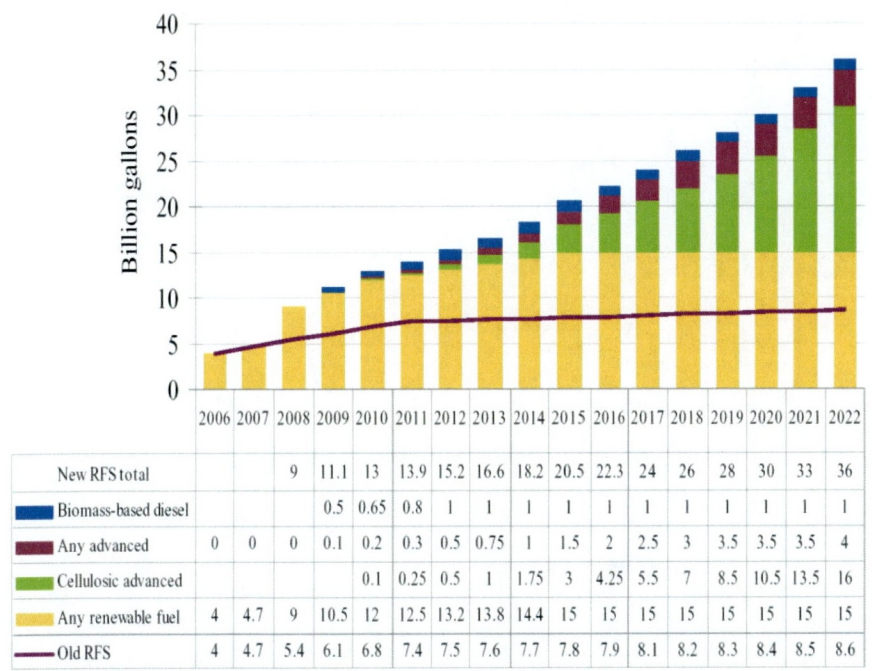

Figure 34. EISA 2007 RFS mandated volumes through 2022

APPENDIX C: FEEDSTOCK AND CONVERSION TECHNOLOGY DETAILS

NREL developed a complete set of plant gate price curves for each technology in each country in the study (Bain, 2007). The conversion processes for ethanol included corn dry mills, sugar cane mills, wheat mills, bio-chemical (B-C) conversion of cellulosic feedstocks, and thermo-chemical (T-C) conversion of cellulosic feedstocks[44]. Representative flow diagrams that show the ethanol yields these conversion processes as well as Fisher-Tropsch catalysis to produce BTL[45] are shown in the following figures. Detailed discussion and analysis of the conversion technologies can be found in the report by Bain.

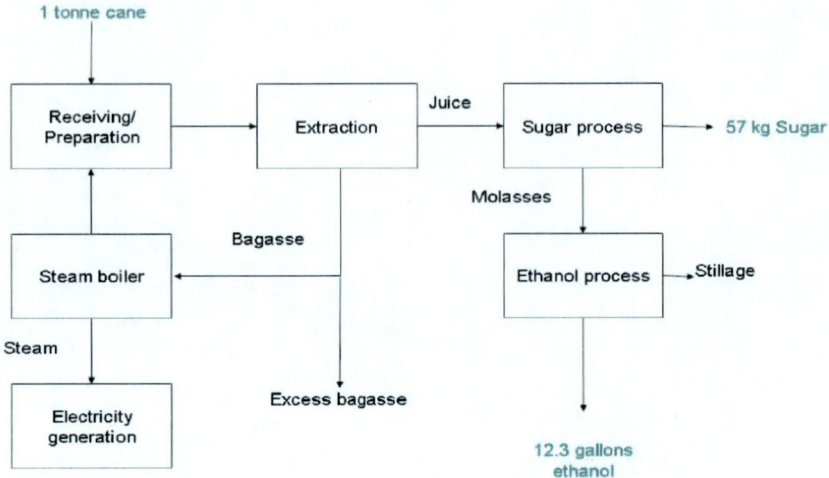

Figure 35. Conversion process to produce sugarcane ethanol

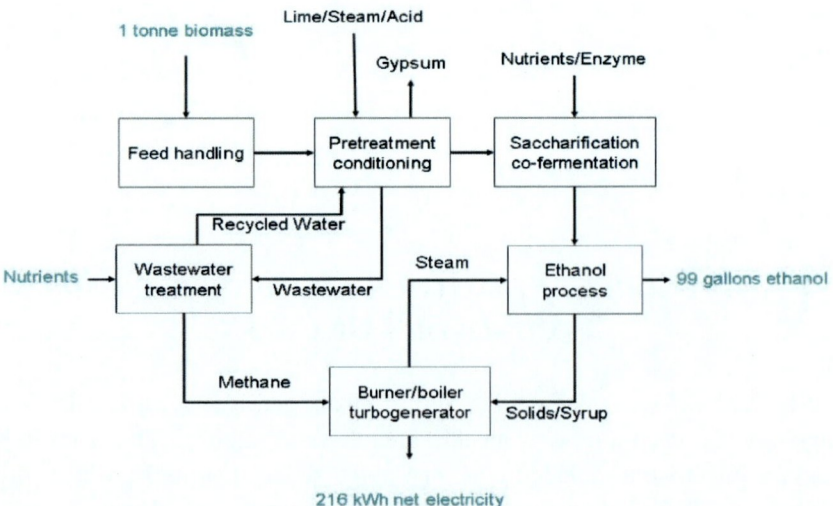

Figure 36. Biochemical conversion process to produce cellulosic ethanol

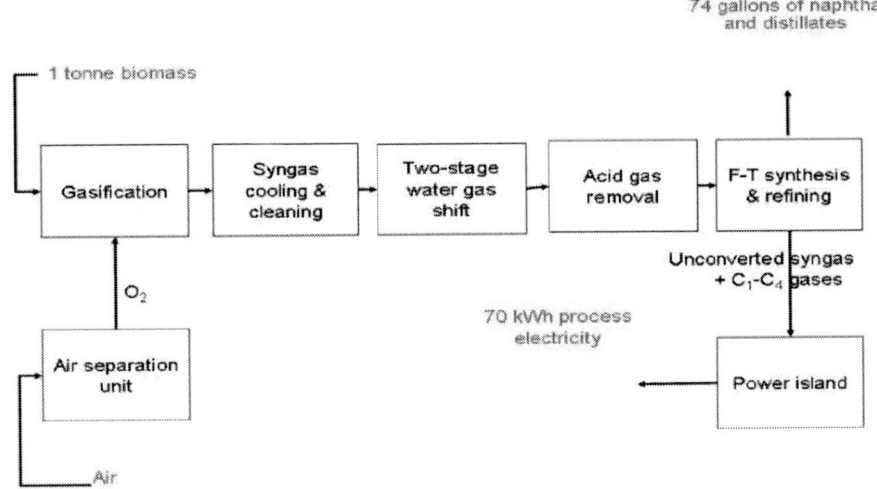

Figure 37. Gasification and Fischer-Tropsch catalysis to produce cellulosic BTL

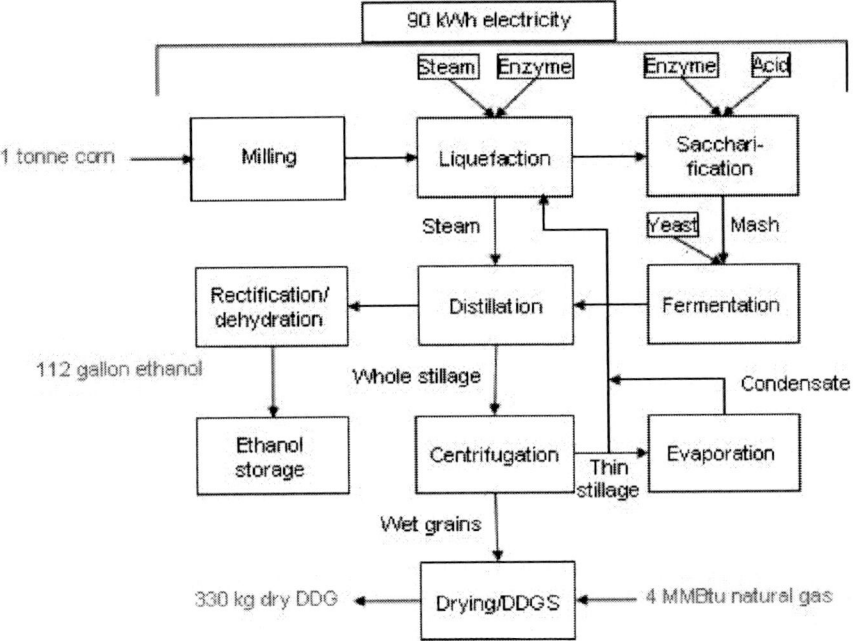

Figure 38. Dry mill conversion process to produce corn ethanol

BIBLIOGRAPHY

Alfstad, T. (2008), *World Biofuels Study- Scenario Analysis of Global Biofuel Markets*, BNL8023 8-2008, May 12, 2008, Brookhaven National Laboratory, Upton, NY.

Armstrong, A.P., Baro, J., Dartoy, J., Groves, A.P., Nikkonen, J., Rickeard, D.J., Thompson, N.D. J-F Larivé (2002), *Energy and Greenhouse Gas Balance of Biofuels for Europe—An Update,* 2002, CONCAWE, Brussels.

Bain, R. (2007), *World Biofuels Assessment: Worldwide Biomass Potential: Technology Characterizations*, Milestone Report NREL/MP-510-42467, December 4, 2007, National Renewable Energy Laboratory, Golden, CO.

Bradley et al. (2007). *Biofuels for Transport: Global Potential and Implications for Energy and Agriculture*, prepared by Worldwatch Institute for the German Ministry of Food, Agriculture and Consumer Protection (BMELV), Earthscan, London.

EIA (2007), *Annual Energy Outlook 2007*, Energy Information Administration, Washington, DC.

EIA (2008a), *Annual Energy Outlook 2008*, Energy Information Administration, Washington, DC.

— (2008b), *This week in Petroleum*, Energy Information Administration, Washington, DC, [cited September 19, 2008].

EISA (2007), U.S. Congress, *Energy Independence and Security Act of 2007*, P.L. 110-140, H.R. 6.

Fargione, J., Jason Hill, David Tilman, Stephen Polasky, & Peter Hawthorne (2008), "Land Clearing and the Biofuel Carbon Debt," *Science, 319*(5867), 1235-1238.

Farm Bill (2008), U.S. Congress, *Food, Conservation, and Energy Act of 2007*, P.L.110-246 H.R. 2419.

Fishbone, L. & Harold Abilock. (1981), "A linear programming model for energy systems analysis: technical description of the BNL version," *Energy Research, 5* (4), p. 353-375.

Hamilton LD, Goldstein., G., Lee, J.C., Marcuse, W., Morris, S.C., Manne, A.S., Wene, C.O. (1992), *MARKAL-MACRO: An overview*, Brookhaven National Laboratory, Upton, NY.

IEA (2007), *World Energy Outlook 2007*, International Energy Agency, Paris, France.

IEA (2006), *Energy Technology Perspectives 2006*, International Energy Agency, Paris, France.

IPCC (2007), *Climate Change 2007: Impacts, Adaptation and Vulnerability. Contribution of Working Group II to the Fourth Assessment Report of the Intergovernmental Panel on Climate Change*, Cambridge University Press, Cambridge, UK, 976pp.

Kline, Keith L., Gbadebo A. Oladosu, Amy K. Wolfe, Robert D. Perlack, Virginia H. Dale, & Matthew McMahon (2007), *Biofuel Feedstock Assessment for Selected Countries*, ORNL Report #ORNL/TM-2007/224, December 3, 2007, Oak Ridge National Laboratory, Oak Ridge, TN.

Larson, E. D. (2008), *"Biofuel production technologies: status, prospects and implications for trade and development,"* United Nations Conference on Trade and Development.

Macedo, I.C., Seabra, J. E. A. & Silva J. E. A. R. (2008), "Green house gases emissions in the production and use of ethanol from sugarcane in Brazil," *The Journal of Biomass and Energy*, doi: 10.101 6/j .biombioe.2007. 12.006 .

Model," submitted to *International Journal of Life Cycle Assessment*, July 20, 2007.

National Ethanol Vehicle Coalition (2008), *E85 Fuel stations*, cited 8 May 2008, available from http://e85vehicles.com/e85-stations.htm.

Peters, S. M. & Timmerhaus K. D. (2003). Plant Design and Economics for Chemical Engineers, McGraw-Hill, NY, NY, ISBN 0-07-049613-7.

President George W. Bush (2007), *State of the Union Address*, Whitehouse, Washington DC.

"Proposal for a Directive of the European Parliament and of the Council on the promotion of the use of energy from renewable sources" {COM(2008) 30 final} {SEC(2008) 57} {SEC(2008) 85} /* COM/2008/0019 final - COD 2008/0016 */ *http://eurlex.europa.eu/LexUriServ/LexUri Serv.do?uri= COM: 2008:0019:FIN:EN:HTML.*

Searchinger, T., Ralph Heimlich, Houghton, R. A., Fengxia Dong, Amani Elobeid, Jacinto Fabiosa, Simla Tokgoz, Dermot Hayes, and Tun-Hsiang Yu (2008), "Use of U.S. Croplands for Biofuels Increases Greenhouse Gases Through Emissions from Land Use Change," *Science, 319,*(5867): 1238-1240.

Sheehan, J., Vince Camobreco, James Duffield, Michael Graboski, and Housein Shapouri (2008), *An Overview of Biodiesel and Petroleum Diesel Life Cycles*, National Renewable Energy Laboratory: Golden, Colorado.

Spath, P., Aden, A., Eggeman, T., Ringer, M., Wallace, B., Jechura J. *(2005). Biomass to Hydrogen Production Detailed Design and Economics Utilizing the Battelle Columbus Laboratory Indirectly-Heated Gasifier,*

NREL Report No. TP-510-37408, National Renewable Energy Laboratory, Golden, CO, 161 pp.

USDA, 2004. *China's Wheat Economy: Current Trends and Prospects for Imports,* prepared by Bryan Lohmar, WHS 04D-01, May 2004, available athttp://www.ers.usda.gov/publications/whs/may04/whs04D01/whs04D01.pdf, [cited July 2007].

Valle-Riestra, J. F. (1983). *Project Evaluation in the Chemical Process Industries,* McGraw-Hill, NY, NY, ISBN 0-07-066840.

Wang, M, Wu, M., Huo, H. & Liu J. (2008), "Life-cycle energy use and greenhouse gas emission implications of Brazilian sugarcane ethanol simulated with the GREET model," International Sugar Journal, September 2008.

Wang, M, Wu, M., Huo, H. & Liu J. (2007), "Well-to-Wheels Energy Use and Greenhouse Gas Emissions of Brazilian Sugarcane Ethanol Production Simulated by Using the GREET

Wang, M., Saricks, C. & Santini D. (1999), *Effects of Fuel Ethanol Use on Fuel-Cycle Energy and Greenhouse Gas Emissions,* Center for Transportation Research, Argonne National Laboratory, Argonne, IL.

Wang, M., Wu, M. & Huo H. (2007), "Life-cycle energy and greenhouse gas emission impacts of different corn ethanol plant types," *Environ. Res. Lett.* 2(April–June 2007), doi: 10.1088/1748-9326/2/2/024001.

White House (2007), *Twenty in Ten: Strengthening America's Energy Security,* available at http://www.whitehouse.gov/ stateoftheunion/ 2007/initiatives/energy

Yacobucci, Brent D. (2008), "Biofuels Incentives: A Summary of Federal Programs," CRS Report for Congress RL33572, July 29, 2008.

Yacobucci, Brent D. & Randy Schnepf (2007). "Selected Issues Related to an Expansion of the Renewable Fuel Standard (RFS)," CRS Report for Congress RL34265, December 3, 2007.

End Notes

[1] EISA is also known as Public Law 110-140.

[2] The 36 B gal per year of renewable fuels required by 2022 include 35 B gal of ethanol-equivalent renewable fuel and 1 B gal of biomass based diesel. Renewable fuels that have more or less energy per gal than ethanol are assumed to be given proportionally more or less credit in meeting this requirement. See Appendix B.

[3] "2008 Farm Bill" refers to the Food, Conservation, and Energy Act of 2008 which was enacted in June 2008. It is also known as P.L.110-246 and includes a cellulosic biofuel tax credit (Sec.15321) of $1.01 per gal for cellulosic biofuel producers through 2012. This tax credit

is part of a biofuel provisions package in the farm bill that includes substantial RD&D funds (including loan guarantees) as mandatory programs, and a reduction in the corn ethanol volumetric ethanol excise tax credit (VEETC).

[4] See http://www1.eere.energy.gov/biomass/

[5] The viability of this scenario is dependent upon resolving issues related to the impact higher blends of ethanol will have on engines and fuel systems. Currently most car manufacturers will only warrant their gasoline engines if they are fuelled with ethanol blends of 10% or less. Blends higher than 10% ethanol are not currently allowed in the U.S.

[6] Both cellulosic ethanol and BTL fuels are qualified cellulosic biofuels as defined by EISA. It is not yet certain whether BTL diesel fuel would also qualify as biomass-based diesel under EISA.

[7] In 2008, the United States imported 72% of the crude oil and petroleum products it consumed. The Energy Information Administration's (EIA) Annual Energy Outlook (AEO) 2008 reference scenario projects that U.S. petroleum imports will increase for the foreseeable future. From a global perspective, energy security is far from assured because most of the world's long-term supplies of accessible crude oil deposits are in one place – the Middle East.

[8] According to the Renewable Fuels Association, in 2007 U.S. ethanol production was 6.5 B gal. Biodiesel production was about 91 million gal in 2005, based on data from the USDA Commodity Credit Corporation which ended its program and its data collection on March 31, 2006.

[9] The countries are Argentina, Brazil, Canada, China, Columbia, India, Mexico, and, collectively, the nations in the Caribbean Basin Initiative.

[10] See Table 2 in the section on Methodology for feedstocks studied in each country (or region).

[11] "2008 Farm Bill" refers to the Food, Conservation, and Energy Act of 2008 which was enacted in June 2008. It is also known as P.L.110-246 and includes a cellulosic biofuel tax credit (Sec.15321) of $1.01 per gal for cellulosic biofuel producers through 2012. This tax credit is part of a biofuel provisions package in the farm bill that includes substantial research, development, and deployment (RD&D) funds (including loan guarantees) as mandatory programs, and a reduction in the corn ethanol volumetric ethanol excise tax credit (VEETC) from $0.51 to $0.45 per gal in 2009-10—which pays for the cellulosic credit as well as the RD&D programs.

[12] Because the MARKAL model works in five year time increments, intermediate year policy changes that came with the 2008 Farm bill—such as the 2009 reduction in the corn ethanol VEETC to $0.45 per gal, and the extension to 2012 of the $0.54 per gal ethanol import tariff cannot be explicitly modeled but are not expected to affect the results significantly.

[13] The viability of this scenario is dependent upon resolving issues related to the impact higher blends of ethanol will have on engines and fuel systems. Currently most car manufacturers will only warrant their gasoline engines if they are fuelled with ethanol blends of 10% or less. Blends higher than 10% ethanol are not currently allowed in the U.S.

[14] This study analyzed the $0.51 per gal tax credit given for blending ethanol into gasoline and the $0.54 per gal import tariff that is charged to ethanol producers outside NAFTA and certain other agreements. These two polices are considered in tandem as the main purpose of the tariff is to cancel out the subsidy for foreign producers. The policies are currently due to expire in 2009 and 2010 but in this scenario an extension is explored.

[15] See Methodology section for further discussion of worldwide biofuels policies.

[16] This study does not model lifecycle greenhouse gas emissions explicitly but does address the associated issues qualitatively.

[17] Existing corn ethanol plants will likely be grandfathered but the rulemaking process is currently ongoing.

[18] A more detailed discussion of the ETP model is presented in the report by Alfstad.

[19] Data sources include U.S. Department of Agriculture (USDA), the Food and Agriculture Organization of the United Nations (FAO), DOE, International Energy Agency (IEA), and other regional and national sources.

[20] Where data were not readily available to develop a meaningful supply curve to update the model, the existing average point estimates in the ETP model were used. (Table 2)

[21] The known cost in one location, generally the U.S. in this study, is adjusted to account for the relative capital and operating costs in each other location. The feedstock cost estimates and tax laws for each country are also incorporated in the result. The actual costs are typically between 90% and 140% of the estimate generated by this type of analysis.

[22] Ethanol by thermochemical conversion is not explicitly separated from ethanol by biochemical conversion in the ETP model because the costs are similar and both produce ethanol.

[23] The ETP model developed by the International Energy Agency is the basis for the DOE's ETP model. The DOE and IEA collaborate to continually refine the ETP model and exchange respective data that define the reference energy system for each of the fifteen world regions: Africa, Australia and New Zealand, Canada, Central and South America, China, Eastern Europe, Former Soviet Union, India, Japan, Mexico, Middle East, Other Asia, South Korea, United States, and Western Europe.

[24] The biomass-based diesel portion of the new RFS is a small share compared to the other renewable fuel type; therefore, this study did not analyze biodiesel market dynamics in detail. The biodiesel tax credit is assumed to continue to 2030 in the model, but it will likely expire at the end of 2008 and not be renewed. This is not expected to change the results of this study significantly because of the small role that biodiesel plays compared to other fuels.

[25] As defined in EISA.

[26] This is extrapolated from a provision of the Food, Conservation, and Energy Act of 2008.

[27] Biofuels mandates and incentives continue to change around the world. This study attempted to use the most recent policies possible.

[28] Each step on the supply curve represents the average production cost for a state or province, while the horizontal length of the step reflects the additional supply projection for that state. Thus, any given point on the curve will represent a cumulative supply for all states producing at or below a corresponding average cost.

[29] This represents total projected cellulosic supplies converted at a rate of 60 gal per dry t.

[30] Since the rate of conversion to fuel (gal per t) varies widely among feedstock types, it is preferable to make any aggregate comparisons in terms of a common energy equivalent. Conversion rates assumed for each feedstock in gal per dry t are 18.2 for sugarcane; 52 for soybeans; 76 for corn; 63 for wheat; 280 for palm oil; and, 60 gal per dry t for cellulosic feedstocks. These conversion rates will vary from the results of the NREL study and those used in the ETP model.

[31] The feedstock assessment's baseline case is defined in the Introduction Section of Kline, 2007.

[32] Brazil, Colombia and Argentina represent about 90% of the total estimated perennial supplies, illustrating the differences in relative scale amongst the countries studied.

[33] Ethanol by thermochemical conversion is not explicitly separated from ethanol by biochemical conversion in the ETP model because the costs are similar and both produce ethanol.

[34] Feedstock price is assume to be at 50% of the 2017 potential supply curve or the single point value if a supply curve is not available.

[35] The model results throughout this discussion are presented in ethanol equivalent volumes.

[36] The 2008 Farm Bill refers to the The Food, Conservation, and Energy Act of 2008 (P.L.110-246) and includes a cellulosic biofuel tax credit (Sec.15321) of $1.01 per gal for cellulosic biofuel producers through 2012. This tax credit is part of a biofuel provisions package in the farm bill that includes substantial RD&D funds (including loan guarantees) as mandatory programs, and a reduction in the corn ethanol volumetric ethanol excise tax credit (VEETC).

[37] U.S. production of cellulosic biofuels in some years is slightly lower in this study compared with the AEO results due to competition with imported cellulosic biofuels, even though the total of domestic production and imports is greater in all years in this study.

[38] Sugar ethanol imports increase in 2025 compared to the reference scenario because domestic cellulosic biofuels are not widely available yet to meet the large increases in the RFS mandate in the 2020 to 2025 timeframe.

[39] The model results throughout this discussion are presented in ethanol equivalent volumes unless otherwise noted.

[40] Here, feedstock availability refers to the physical presence of biomass resources in the region. Availability at conversion plants depends on the ability to harvest, collect, and transport the feedstocks, and this is treated as an infrastructure constraint in this study.

[41] This is how the ETP model behaves; strictly enforcing mandates. In the real world lawmakers and regulators would likely issue waivers, adjust the regulations, or otherwise accommodate markets if the mandates are deemed to have unacceptable adverse impacts.

[42] It is not yet clear how the 1 B gal biomass-based diesel requirement will be calculated – in gal of ethanol equivalent or gal of diesel equivalent.

[43] There are further requirements and qualifications for biomass-based diesel as defined in EISA. It is not yet clear whether Fischer-Tropsch BTL diesel would qualify as both a cellulosic biofuel and a biomass-based diesel. Biomass-based diesels are mainly limited to fatty acid methyl esters (FAME) made from oil.

[44] Ethanol by thermochemical conversion is not explicitly separated from ethanol by biochemical conversion in the integrated assessment because the costs are similar and both produce ethanol.

[45] Conversion costs for cellulosic BTL conversion are from the ETP model.

CHAPTER SOURCES

The following chapters have been previously published:

Chapter 1 – This is an edited, reformatted and augmented version of a United States Congressional Research Service publication, Report Order Code RS21930, dated March 18, 2008.

Chapter 2 – This is an edited, reformatted and augmented version of a United States Congressional Research Service publication, Report Order Code R40155, dated January 23, 2009.

Chapter 3 – This is an edited, reformatted and augmented version of a United States Department of Energy, Office of Policy Analysis, Office of Policy and International Affairs publication, dated September 15, 2008.

INDEX

A

accounting, 29, 67
ACE, 7
acid, xii, 39, 63, 107
adaptability, 29
adjustment, 84
AEI, 34, 36
aggregate supply, 55, 60
aggregate supply curve, 55
agricultural, x, xi, 2, 10, 11, 12, 13, 14, 17, 20, 23, 25, 26, 29, 33, 34, 58, 62, 63
agricultural commodities, x, 2
agricultural crop, 25, 62
agricultural market, 26
agricultural residue, 12, 34
agricultural sector, 29
agriculture, 25
air, 32
air quality, 32
alcohol, 2, 7
algae, 24, 49
alternative, x, 2, 9, 10, 11, 15, 19, 22, 29, 33, 43, 46, 47, 53
alternative energy, 15, 33
ammonia, 54
AMS, 18
animals, 27
appendix, 80

application, 77
argument, 21, 26, 86
asia, 63, 80, 88, 106
asian, 48, 80
asian countries, 80
assessment, 42, 85, 106, 107
assumptions, 41, 44, 48, 63, 69, 80, 84, 94
atmosphere, 47
authority, 69
automakers, 31
availability, ix, xii, 29, 32, 34, 38, 39, 48, 50, 54, 61, 62, 65, 67, 69, 75, 76, 77, 78, 80, 81, 84, 85, 86, 107

B

back, 16, 48, 91
barges, 30
barrier, 3, 12, 19, 32
beef, 27
behavior, 46, 55, 82
benefits, 15, 21, 23, 25, 29, 33, 34, 41, 46, 47, 73, 96
binding, 48
biodegradable, 15
biodiesel, xi, xii, 10, 11, 12, 13, 14, 20, 22, 24, 33, 39, 56, 62, 69, 85, 89, 90, 106
biodiversity, 42, 48

biomass, xi, xii, 10, 13, 20, 21, 22, 24, 30, 34, 38, 39, 43, 47, 49, 52, 55, 62, 67, 72, 80, 82, 89, 98, 104, 105, 106, 107
blends, 31, 32, 63, 72, 82, 105
boilers, 62
Brazilian, 3, 11, 12, 20, 33, 48, 57, 59, 67, 68, 75, 77, 87, 104
broilers, 18
burn, 48

C

candidates, 63
capital cost, 41, 69, 80
capital intensive, 64
carbon, 15, 25, 40, 43, 44, 45, 47, 48, 81, 82, 86, 87, 95
carbon credits, 15
carbon dioxide, 45, 81
carbon emissions, 41, 43
Caribbean Basin Initiative (CBI), vii, ix, x, 1, 2, 4, 7, 8, 12, 33, 105
Caribbean countries, 13
Caribbean nations, 96
catalysis, 52, 68, 72, 99, 101
cattle, 26, 27
cellulose, x, 9, 13, 20, 21, 49, 63
cellulosic ethanol, xii, 19, 22, 34, 39, 40, 43, 44, 46, 55, 64, 65, 68, 69, 72, 73, 75, 78, 81, 86, 91, 94, 95, 100, 105
cement, 54
Central America, ix, x, 1, 2, 4
Central America Free Trade Agreement (CAFTA), ix, x, 2, 4, 8
certification, 71, 75, 78
channels, 26
chemicals, 20
Clean Air Act, 5, 98
climate change, x, 9, 10
coal, 11, 24
combustion, 2, 22, 47, 62
commercialization, 19, 42

commodity, 5, 14, 18, 26, 28, 33, 91
Commodity Credit Corporation, 105
competition, 27, 39, 40, 43, 48, 78, 86, 87, 90, 92, 94, 107
competitor, 3
compliance, 41, 91, 98
components, 15, 31, 52, 53
concentration, 27
confidence, 95
configuration, 54
conflict, 33
congress, vi, 4, 6, 11, 19, 33, 41, 102, 104
consensus, 49
conservation, 14, 33
constraints, 13, 24, 40, 41, 42, 54, 55, 71, 72, 74, 75, 78, 86, 90, 93, 94, 95
construction, 8, 16, 29, 95
Consumer Price Index (CPI), 26, 28, 35
consumers, 28, 75, 82, 92, 93, 94
consumption, ix, 1, 3, 5, 6, 8, 11, 12, 22, 23, 27, 28, 32, 34, 35, 40, 42, 66, 67, 73, 78, 97
contract prices, 83
contractors, 55
contracts, 17
conversion, xi, 19, 21, 35, 38, 41, 42, 44, 47, 50, 52, 53, 55, 58, 63, 64, 68, 72, 99, 100, 101, 106, 107
conversion rate, 35, 106
corn-starch ethanol, 20, 23
Corporate Average Fuel Economy, 98
corrosion, 30, 31
cost curve, 55
cost minimization, 53
cost saving, 20
cost-effective, 46, 53, 64
costs, xi, 3, 11, 14, 19, 20, 24, 26, 27, 28, 31, 32, 33, 34, 38, 41, 42, 44, 46, 50, 53, 55, 57, 62, 63, 64, 69, 72, 77, 78, 80, 83, 91, 94, 106, 107
covering, ix, xii, 38, 55

Index

credit, xii, 17, 34, 38, 40, 41, 43, 44, 49, 56, 68, 69, 73, 74, 77, 80, 81, 90, 91, 94, 95, 96, 104, 105, 106
criticism, 15
crop production, 23, 25
crop residues, 34, 62
crops, x, xi, 9, 10, 11, 12, 20, 21, 25, 26, 34, 42, 47, 49, 50, 51, 52, 57, 58, 61, 62, 63, 64, 85
CRS, xi, 7, 8, 10, 13, 17, 18, 19, 33, 34, 35, 104
crude oil, 35, 44, 47, 82, 83, 105
cultivation, 24, 47, 57
customers, 72, 92

D

dairy, 26, 27
data collection, 105
database, 36, 55
debt, 19, 47, 48
decision makers, 54
definition, 24, 33, 34
deforestation, 47, 48
degradation, 33
dehydration, 3, 4
delivery, 22, 49, 53
Department of Agriculture, 106
Department of Energy (DOE), xi, 34, 38, 109
deposits, 105
derivatives, 5
developing countries, 28
developing nations, 49
diesel, xii, 12, 13, 24, 30, 31, 32, 38, 40, 47, 49, 52, 63, 67, 78, 91, 94, 104, 105, 106, 107
diesel fuel, 13, 67, 105
diminishing returns, 95
discount rate, 55
distillates, 72
distortions, 27

distribution, 21, 30, 31, 40, 49, 53, 63, 72, 75, 82, 90, 92, 94, 95
diversity, 42
domestic demand, 57, 59, 61
drought, 23
drying, 48
duties, ix, 1, 5
duty free, 12
duty-free treatment, ix, 2

E

Eastern Europe, 106
ecological, 42
economic activity, 29
economic competitiveness, xi, 38, 42
economic growth, 26
economic incentives, 24, 39, 81, 91, 97, 98
economic rent, 82
Economic Research Service, 35, 36
economics, 24, 26, 40, 74, 75, 81, 87, 95
economies of scale, 27
ecosystem, 48
egg, 28
electricity, 21, 62
emission, 47, 104
employment, 25, 29, 49
encouragement, 97
energy consumption, 22
energy efficiency, 97, 98
Energy Efficiency and Renewable Energy, 37
Energy Independence and Security Act, ix, x, xi, 1, 5, 9, 11, 14, 33, 38, 42, 97, 102
Energy Information Administration (EIA), 33, 34, 42, 102, 105
energy markets, 53
Energy Policy Act, ix, x, 1, 5, 9, 13, 14, 98
Energy Policy Act of 2005, ix, x, 1, 5, 9, 13, 14, 98
energy supply, xi, 10, 12, 13
engines, 82, 105

environment, 12, 41, 42, 43, 77, 94
environmental degradation, 33
environmental effects, xi, 10, 20
environmental impact, 19
Environmental Protection Agency (EPA), xii, 32, 35, 38, 49, 69, 99
enzymatic, 19
epoxy, 31
equilibrium, 53
erosion, 21, 25
esters, 39, 63, 107
estimating, 42, 47
ethyl alcohol, 7
European Parliament, 103
European Union, 46
Europeans, 46
excise tax, 7, 105, 106
exercise, 84, 92
expansions, 17, 25
expenditures, 28
exporter, 7, 29, 66
Export-Import Bank, 8
exports, 5, 29, 39, 42, 43, 53, 58, 89
extraction, 49, 53, 55

F

FAO, 106
Farm Bill, xii, 18, 38, 41, 43, 49, 69, 80, 81, 102, 104, 105, 106
farm land, 47
farmers, 2, 18, 29, 49, 82
farming, 50
fatty acid, xii, 39, 107
Federal Highway Administration, 36
Federal Register, 8
feedback, 86
fermentation, 2, 52
fertility, 21
fertilizer, 22, 23, 25, 43, 47
fiber, 51, 57, 62
field crops, 25

first generation, 86
Fischer-Tropsch synthesis, 64
flex, 93
flexibility, 28
flow, xi, 38, 53, 54, 91, 99
food, xi, 10, 11, 12, 14, 16, 17, 20, 26, 27, 28, 39, 42, 43, 47, 49, 51, 57, 58, 61, 77, 80, 81, 85
food products, 26, 27
food stamp, 28
foreign producer, 96, 105
forestry, 62, 63
fossil, 7, 15, 20, 22, 33, 40, 48, 98
fossil fuel, 7, 15, 20, 22, 40, 48, 98
fuel cycle, 22
fuel efficiency, 93, 98
fuel type, 48, 96, 106
fuelwood, 63
funds, 30, 34, 75, 81, 105, 106
futures, 17

G

gas, xii, 11, 12, 14, 15, 20, 22, 23, 24, 33, 39, 42, 49, 98, 104, 105
gasoline, ix, x, 1, 2, 3, 5, 6, 9, 10, 12, 15, 17, 19, 21, 22, 23, 24, 25, 30, 31, 32, 33, 34, 35, 40, 44, 49, 56, 71, 72, 78, 80, 82, 91, 92, 93, 95, 99, 105
GDP, 29
General Accounting Office, 7
generation, 42, 49, 50, 67, 86
global climate change, x, 9, 10
global markets, 58
global warming, 49
globalization, 36
goals, xi, 10, 33, 39
government, vi, xi, 10, 17, 19, 28, 29, 30, 46, 50, 61, 95
government intervention, xi, 10
grain, xi, 10, 11, 12, 26, 28, 29, 41, 43, 67, 73, 78, 85, 86, 87, 90, 91, 94

Index 115

grasses, 12, 20, 48
greenhouse, xii, 14, 15, 20, 22, 24, 33, 39, 42, 49, 98, 104, 105
greenhouse gas (GHG), xii, 14, 15, 20, 22, 24, 33, 39, 42, 49, 98, 104, 105
groups, 28, 34
growth, 5, 11, 16, 17, 22, 26, 29, 30, 39, 41, 43, 49, 51, 56, 57, 58, 59, 61, 68, 69, 73, 77, 78, 85, 86, 87, 89, 95
growth rate, 57, 58, 61, 68, 85, 95
guidance, 37

H

handling, 53
harmful effects, 26
harvest, 17, 20, 47, 107
heart, 3
heat, 62
heating, 15, 16
heating oil, 15, 16
higher quality, 83
home heating oil, 16
horizon, 53
house, 8, 35, 104
household, 28, 98
humanitarian, 28
hydro, 52
hypothesis, 71

I

implementation, xii, 14, 38
import prices, 83
importer, 23
imports, ix, x, 1, 2, 3, 4, 5, 6, 7, 12, 15, 20, 21, 23, 24, 28, 31, 33, 34, 35, 40, 41, 42, 43, 44, 46, 49, 53, 69, 71, 73, 78, 80, 90, 92, 95, 96, 97, 105, 107
incentive, ix, x, 1, 2, 3, 24, 25, 43, 56, 67, 73, 74, 93, 96
income, 28, 29

independence, 34
Indian, 66
indication, 86
industry, x, xi, 6, 7, 10, 16, 17, 29, 30, 31, 33, 34, 39, 40, 77, 94
inelastic, 41, 71, 77, 78, 86
inflation, 26, 82, 95
infrastructure, xi, 10, 13, 14, 19, 30, 31, 32, 39, 40, 41, 46, 50, 53, 71, 72, 74, 75, 78, 82, 86, 90, 92, 93, 94, 95, 97, 107
innovation, 39
insight, 41
interaction, 55, 78
Intergovernmental Panel on Climate Change (IPCC), 41, 103
International Energy Agency (IEA), 102, 106
international markets, 28, 40, 47, 94
International Monetary Fund, 36
International Trade, 3, 6, 8, 46
International Trade Commission, 3, 6, 8
intervention, xi, 10
invasive, 21
invasive species, 21
inventories, 28
investment, 14, 15, 30, 39, 44, 47, 55, 81, 95
investment capital, 14
investors, 47, 95
isolation, 84

J

jet fuel, 5
jobs, 29

L

labor, 43, 55
land, xii, 11, 14, 20, 25, 39, 42, 43, 47, 48, 49, 51, 58, 59, 62
land tenure, 47

land use, xii, 14, 20, 25, 39, 42, 47, 48, 49, 51
large-scale, 25
law, 7, 11, 40, 77, 97, 98
learning, 44, 68, 69, 80, 81
legislation, 4, 33, 92
lenders, 17, 40, 92, 95, 98
lending, 3, 44, 56
Life Cycle Assessment, 103
lifecycle, 15, 22, 24, 25, 33, 35, 47, 48, 49, 73, 98, 105
light trucks, 31
limitation, 39, 75, 95
linear, 102
linear programming, 102
liquid fuels, xii, 39
liquids, 43, 52
livestock, 11, 26, 27
loan guarantees, 105, 106
location, 21, 106
long distance, 27
losses, 48
lower prices, 94
lower-income, 28
low-income, 28

M

machinery, 21
maintenance, 53
mandates, 40, 41, 42, 43, 46, 56, 68, 69, 71, 72, 75, 78, 87, 88, 89, 91, 93, 94, 98, 106, 107
manufacturing, 5, 6, 7
market penetration, 46, 55
market prices, 28
market share, 39, 46, 55, 56, 67, 85, 86, 87, 92
market value, 91
marketplace, 26, 27, 68
markets, 3, 15, 16, 21, 26, 27, 28, 29, 39, 40, 43, 46, 47, 53, 56, 58, 59, 71, 77, 78, 82, 84, 86, 87, 88, 91, 92, 94, 95, 96, 97, 107
meat, 27
media, 35
metals, 54
methanol, 13
microbial, 19
Middle East, 105, 106
million barrels per day, 21
mining, 53
missions, 14, 15, 20, 22, 24, 25, 33, 42, 47, 49, 98, 103, 105
models, 53, 54, 69
motivation, 3
movement, xi, 10, 11
MTBE, 3, 7
municipal solid waste, 31

N

NAFTA, 96, 105
nation, 3
national product, 16
national security, 42
natural, 11, 12, 22, 23, 24, 54
natural gas, 11, 12, 22, 23, 24
network, 46, 53
nitrogen, 22
nodes, 53
non-native, 21
normal, 24, 55
nutrition, 26
NYMEX, 83

O

octane, 2
OECD, 36, 44
oil palm, 63
oil production, 61
oils, 12, 13, 47
online, 35

opportunity costs, 19
optimization, xi, 38
Organization of the Petroleum Exporting Countries (OPEC), 82
output method, 29
ozone, 5

P

palm oil, 48, 61, 62, 106
Parliament, 103
passenger, 22
pasture, 58
pathways, 53
peatland, 48
pesticides, 21
petroleum, x, 5, 6, 9, 10, 11, 14, 15, 22, 23, 24, 30, 31, 35, 42, 43, 72, 93, 94, 98, 105
petroleum products, 6, 30, 105
photosynthesis, 21
pipelines, 30, 31, 36, 72
planning, 47, 53, 54
plants, 2, 4, 19, 21, 25, 29, 34, 52, 61, 62, 81, 105, 107
play, ix, 1, 11, 13, 31, 68
point of origin, x, 2
policymakers, 15, 34, 39
political stability, 80
pollutant, 14
pollution, 5, 47
poor, 61, 74
pork, 27, 28
poultry, 26, 27
power, 20
premium, 40, 55, 91, 92, 93, 94
press, 83
pressure, 17, 46, 48
price changes, 83
price effect, 27
price signals, 40, 86, 87, 91
private, 49, 81

producers, xii, 5, 27, 34, 38, 39, 46, 56, 59, 61, 73, 80, 82, 90, 91, 92, 95, 96, 97, 104, 105, 106
production costs, 3, 19, 24, 46, 57
production technology, 58
productivity, 62, 63
profit margin, 27
profitability, 3, 19, 62
profits, 26, 27, 91
program, 17, 28, 105
protein, 18, 27
PSD, 36
public, 14, 15, 48, 49, 81
public interest, 14
public policy, 14, 15
pulp, 54
pumps, 31, 32, 47
pyrolysis, 52

Q

qualifications, 107

R

rail, 30
rain, 58
rainforest, 48
range, ix, xii, 17, 18, 24, 38, 40, 42, 46, 59, 61, 62, 64, 65, 83, 84, 98
raw material, 30
real terms, 95
reality, 19, 46
recession, 17, 33
reconcile, 21
recovery, 63
refineries, 62, 92, 98
refining, 63
regional, 25, 27, 55, 106
regulations, ix, xii, 17, 38, 107
regulators, 107
relationship, 95

relative prices, 61
renewable energy, 97
Renewable Fuel Standard (RFS), vii, xi, xii, 9, 10, 14, 38, 42, 81, 98, 104
renewable resource, 15
rent, 82, 83
reprocessing, 4
Research and Development, 36
residues, 12, 34, 62, 63
resource availability, 54, 85
resources, 15, 39, 61, 107
retail, 26, 27, 31, 32
returns, 95
revenue, 46
rice, ix, xii, 6, 29, 38, 40, 45, 66, 73, 77, 78, 79, 82, 86, 87, 91, 94, 95, 96, 99
risk, 14, 28, 77, 91, 94
runoff, 21
rural, x, 9, 10, 25, 29, 30
rural areas, 30

S

sales, 29, 72, 87, 93, 94, 95, 96
savings, 20, 48
scarcity, 62
Scenario Analysis, 39, 81, 102
school, 28
security, 14, 23, 24, 28, 33, 34, 41, 42, 47, 105
seeds, 33, 86
senate, 8
sensitivity, 85
services, vi, 54
shape, 86
shares, 46, 58
shipping, 30, 31
short run, 28
short supply, 90
signals, 40, 58, 86, 87, 91
single market, 46
skeptics, 18

SNAP, 28
soft drinks, 27
soil, 21, 25, 34, 47
soil erosion, 21, 25
solid waste, 31
South America, 39, 42, 55, 66, 88, 89, 106
Southeast Asia, 48, 63
soy, 59
soybean, xi, 10, 11, 25, 27, 28, 58, 59, 61, 106
species, 21, 27, 62
speed, 19
stability, ix, 1, 23, 80
stages, 49, 55, 81, 92
stakeholders, 6
standards, 39, 88, 98
starch, x, 5, 9, 11, 14, 15, 16, 20, 23, 33, 36, 42, 43, 51, 52, 98
State of the Union, 11, 103
stock, 48, 54, 55
storage, 30, 32
strain, 30, 31
Strategic Petroleum Reserve, iii
strategies, 33
strength, 95
subsidy, 40, 44, 46, 81, 82, 90, 91, 92, 93, 94, 105
substitutes, 13, 24, 29, 30, 31, 94
sugar, xii, 5, 11, 12, 13, 14, 20, 24, 33, 39, 41, 58, 62, 64, 67, 69, 72, 73, 75, 77, 78, 85, 86, 90, 91, 94, 99
sugarcane, x, 2, 4, 5, 9, 12, 22, 24, 33, 41, 48, 57, 58, 60, 61, 62, 64, 71, 75, 77, 78, 81, 86, 87, 99, 100, 103, 104, 106
supervision, 37
suppliers, 75, 77, 94
supply curve, 41, 50, 51, 52, 55, 56, 71, 75, 77, 78, 82, 84, 85, 86, 106
surplus, 39
sustainability, 48
switching, 82
synthesis, 63, 64
systems, 31, 39, 82, 102, 105

# Index

T

tanks, 30, 31, 32
targets, 47, 66, 91, 92, 94, 97
tariff, ix, 1, 3, 7, 12, 33, 44, 56, 73, 74, 80, 96, 105
tax credit, ix, xii, 17, 34, 38, 40, 41, 43, 44, 49, 56, 68, 69, 80, 81, 90, 94, 95, 96, 104, 105, 106
tax exemptions, 46, 97
tax incentive, x, 2, 3, 7, 56
taxes, 46, 56, 97, 98
taxpayers, 75
technological progress, 53
technology, xi, 15, 19, 31, 38, 39, 41, 43, 44, 46, 50, 52, 53, 54, 55, 57, 58, 64, 68, 69, 73, 78, 80, 81, 86, 90, 94, 99
technology transfer, 41
tenure, 47
terminals, 30
threats, 23, 28, 41
time, 3, 22, 34, 39, 53, 56, 65, 68, 75, 90, 95, 105
time increment, 105
time periods, 65, 90
total energy, 22
total product, 51
toxic, 15
trade, xi, 15, 21, 24, 28, 29, 38, 39, 42, 44, 46, 54, 77, 78, 89, 91, 96, 103
Trade Act, 4
trade-off, 21
trading, 17
trans, 26
transesterification, 52
transfer, 39, 41, 69, 75, 80
transition, 47, 62
transport, 22, 30, 48, 53, 72, 91, 107
transport costs, 91
transportation, ix, x, 1, 9, 10, 12, 15, 22, 30, 31, 52, 63, 68, 98

treasury, 90
trees, 12, 20
trucks, 30, 31, 32, 72, 93

U

U.S. Agency for International Development, 37
U.S. Department of Agriculture (USDA), 16, 17, 18, 21, 26, 34, 35, 36, 104, 105, 106
U.S. economy, 29
U.S. Export-Import Bank, 8
U.S. Treasury, 90
uncertainty, 19, 46, 47, 58, 84
United Nations, 103, 106
upload, 36

V

values, 44, 49, 84, 95
variable costs, 55
variation, 43, 50, 51, 52, 64, 88, 94
vegetable oil, 13
vehicles, xii, 12, 31, 32, 38, 44, 56, 72, 93
venture capital, 15

W

wastes, 49, 62, 63
water, 2, 30, 43, 47
Western Europe, 46, 66, 106
wheat, 26, 28, 61, 63, 69, 99, 106
White House, 104
wholesale, 3, 18, 24, 95, 99
wood, 62
World Resources Institute, 48

Y

yield, 35, 46, 51, 57, 58